Lecture Notes in Mathematics

Edited by A. Dold and B. Eckmann

754

Olav Arnfinn Laudal

Formal Moduli of Algebraic Structures

Springer-Verlag
Berlin Heidelberg New York 1979

Author

Olav Arnfinn Laudal
Matematisk Institutt
Universitetet i Oslo
Postboks 1053
Blindern-Oslo 3
Norway

AMS Subject Classifications (1970): 13D10, 14D15, 14D20, 14F10, 14F99, 18H20, 55G30

ISBN 3-540-09702-3 Springer-Verlag Berlin Heidelberg New York
ISBN 0-387-09702-3 Springer-Verlag New York Heidelberg Berlin

Printing and binding: Beltz Offsetdruck, Hemsbach/Bergstr.
2141/3140-543210

Contents

Introduction. The following pages contain notes of a series of lectures given at the University of Oslo during the year 1974-75.

The subject was deformation theory, and in particular the study of the hulls of some deformation functors encountered in algebraic geometry. The lectures were based upon work done by the author from 1969 to 1975.

Most of the results presented here may be found in two preprints published by the Institute of Mathematics at the University of Oslo, the first in 1971 and the second in 1975, see [La 4] and [La 5].

The main goal of the lectures was the proof of the structure theorem (4.2.4). The first thing

to do was therefore to construct a cohomology theory for categories of morphisms of algebras, and establish an obstruction calculus suitable for our purpose.

The developments of these elements of the theory has a rich history, the details of which I don't feel competent to write.

Nevertheless I think it is proper to mention a few steps and some names - and their relation to the results of these notes.

Inspired, I believe, by results of Kodaira-Spencer and Grothendieck, Schlessinger and Lichtenbaum defined in [Li] a cotangent complex good enough to enable them to prove the first nontrivial theorems relating deformation theory to the cohomology of algebras.

Later André [An] and Quillen [Qu] defined the correct cotangent complex, using quite different technics.

The approach of Quillen was then extended by Illusie [Il] to yield a global theory working nicely for any topos.

Independently I had been working on an extension of the method of André, based on the study of the inductive and projective limit functors on small categories (see [La 1] and [La 2]). My results appeared in the preprint [La 4].

Therefore, at the time of my lecture there existed a well developed cohomology theory and a corresponding obstruction calculus.

I chose to carry out the construction of the global cohomology theory along the lines of [La 4] and this construction occupies the 3 first chapters of these notes. Many of the results found there are therefore not new. Some will, properly translated into the language of Illusie be found in his Springer Lecture Notes, others may be deduced from his general theorems.

The main results of this paper have the following corollaries,

(3.2): Given any S-scheme X and any quasicoherent 0_X-Module M, there exist cohomology groups

$$A^n(S,X;M) \qquad n \geq 0$$

the abutment of a spectral sequence given by the term $E_2^{p,q} = H^p(X,\underline{A}^q(M))$, where the sheaf $\underline{A}^q(M)$ is an 0_X-Module defined by $\underline{A}^q(M)(U) = H^q(S,A;M(U))$ whenever $U = \mathrm{Spec}(A)$ is an affine open subset of X, the last cohomology being that of André.

(3.2): Given any morphism of finite type of S-schemes $f: X \to Y$, and any quasicoherent 0_X-Module M, there exist cohomology groups

$$A^n(S,f;M) \qquad n \geq 0$$

the abutment of a spectral sequence given by the term

$E^{p,q} = H^p(Y,\underline{A}^q(f;M))$, where the sheaf $\underline{A}^q(f;M)$ is an 0_Y-Module defined by $\underline{A}^q(f;M)(V) = A^q(B,f^{-1}(V);M)$ whenever $V = \mathrm{Spec}(B)$ is an open affine subset of Y.

(3.2): Let Z be a locally closed subscheme of the S-scheme X, and let M be any 0_X-Module. Then there are cohomology groups

$$A^n_Z(S,X;M) \qquad n \geq 0 ,$$

the abutment of a spectral sequence given by the term $E_2^{p,q} = A^p(S,X;\underline{H}^q_Z(M))$. Moreover there is a long exact sequence

$$\to A^{n-1}_Z(S,X;M) \to A^{n-1}(S,X;M) \to A^{n-1}(S,X-Z;M)$$

$$\to A^n_Z(S,X;M) \to \cdots$$

(3.3): Given any morphism of S-schemes $f: X \to Y$, and any 0_X-Module M there is a long exact sequence

$$\to A^{n-1}(S,f;M) \to A^{n-1}(S,X;M) \to A^{n-1}(S,Y;R^{\cdot}f_*M)$$

$$\to A^n(S,f;M) \to \cdots$$

(4.1): (4.1.11) and (4.1.17)

Let $\pi: R \to S$ be any surjective homomorphism of rings. Suppose $(\ker\pi)^2 = 0$ and consider a morphism of S-schemes $f: X \to Y$. Then there exists an obstruction element $o(f,\pi) \in A^2(S,f;0_X \otimes_S \ker\pi)$, such that $o(f,\pi) = o$ is a necessary and sufficient condition for the existence of a deformation of f to R (see definitions (4.1.)). The set of such deformations modulo an obvious equivalence relation, is a principal homogenous space over $A^1(S,f;0_X \otimes_S \ker\pi)$.

(5.1.1) Let k be any field, and let $f: X \to Y$ be any morphism of algebraic k-schemes. Then the infinitesimal deformation functor of f has a hull H characterized in the following way: Let T^i denote the completion of the symmetric k-algebra on the (topological) k-dual of $A^i(k,f;\mathcal{O}_X)$ (see (4.2) for definitions), then there exists a morphism of complete k-algebras

$$o : T^2 \to T^1$$

with the following properties:

(i) $o(\underline{m}_{T^2}) \subseteq (\underline{m}_{T^1})^2$

(ii) o is essentially determined by the Massey products of A^*.

(iii) the leading term of o (the primary obstruction) is unique.

(iv) $H = T^1 \hat{\otimes}_{T^2} k$.

(5.2.3) Let Z be a locally closed subscheme of the algebraic k-scheme X and let M be any $\mathcal{O}_X$-Module. Suppose $\inf_{z \in Z} \{\mathrm{depth}_Z M\} \geq n+2$ then

$$A^p(k,X;M) \to A^p(k,X-Z;M)$$

is an isomorphism for $p \leq n$.

(5.2.10) Let X be any closed subscheme of the algebraic k-scheme Y. Suppose X is locally a complete intersection of Y, and suppose $\underline{\mathrm{Hilb}}_Y$ is representable. Then there exists a homomorphism of complete local k-algebras

$$o : \mathrm{Sym}_k(H^1(X,N_{X/Y})^*)^\wedge \to \mathrm{Sym}_k(H^0(X,N_{X/Y})^*)^\wedge$$

such that

$$\hat{O}_{Hilb_Y,\{X\}} \simeq Sym_k(H^0(X,N_{X/Y})^*)\hat{} \;\hat{\otimes}_k\; Sym_k(H^1(X,N_{X/Y})^*)\hat{}$$

where $N_{X/Y}$ is the normal bundle of X in Y.

Notations: Let $\cdot \underset{\varphi}{\to} \cdot \underset{\psi}{\to} \cdot$ be two composable morphisms in some category. We shall denote by $\varphi\psi$ the composition of φ and ψ.

N.B. To avoid set theoretical difficulties we shall assume that all constructions involving categories, sets, etc. take place in a fixed universe. No attempt is made to prove that the results emerging from these constructions are independent of the choice of this universe.

However, this seems rather obvious, see the corresponding discussions in [An].

Acknowledgments. I should like to express my sympathy and my warmest thanks to the audience of my lectures from which these notes are taken. I have profited a lot on discussions with Geir Ellingsrud, Jan Kleppe and Stein Arild Strømme.

The Norwegian Research Council for Science and the Humanities (NAVF) has given its support through the contract nr : D.00.01-37.

Finally, the seemingly infinite patience of Randi Møller, in typing the manuscript, was crucial.

Chapter 1.

Sections of functors:

In this chapter we shall be concerned with the following general problem:

Given functors of small categories

$$\begin{array}{ccc} & & \underline{C} \\ & & \downarrow \pi \\ \underline{e} & \underset{\rho}{\rightarrow} & \underline{c} \end{array}$$

When will there exist a section i.e. a functor $\sigma : \underline{e} \rightarrow \underline{C}$ such that $\sigma \circ \pi = \rho$?

We shall show that under suitable conditions there is a sequence of natural cohomology groups $H^i(\underline{e}, Der_\pi)$ and an obstruction $o \in H^2(\underline{e}, Der_\pi)$ such that a section exists iff $o = 0$. The set of sections modulo an obvious equivalence relation, if non-empty, is a principal homogenous fiberspace over $H^1(\underline{c}, Der_\pi)$.

(1.1) Derivation functors associated to a functor.

Let $\pi : \underline{C} \rightarrow \underline{c}$ be a functor of small categories. We shall consider the category

$$\underline{\text{Mor } \underline{c}} \; ,$$

for which

1. The objects are the morphisms of $\underline{c}$.
2. If φ, φ' are objects in $\underline{\text{Mor } \underline{c}}$ then the set of morphisms $\text{Mor}(\varphi, \varphi')$ is the set of commutative diagrams

$$\begin{array}{ccc} * & \xleftarrow{\psi} & * \\ \varphi \downarrow & & \downarrow \varphi' \\ * & \xrightarrow[\psi']{} & * \end{array}$$

We write $(\psi,\psi'):\varphi \to \varphi'$ for such a morphism.

Let $\varphi \in \underline{\text{Mor}}\ \underline{c}$ be an object (i.e. a morphism of $\underline{c}$) and let $\pi^{-1}(\varphi) = \{\lambda \in \underline{\text{Mor}}\ \underline{C} \mid \pi(\lambda) = \varphi\}$.

If φ_1 and φ_2 are morphisms in $\underline{c}$ which can be composed, then we have a partially defined map:

$$m : \pi^{-1}(\varphi_1) \times \pi^{-1}(\varphi_2) \to \pi^{-1}(\varphi_1 \circ \varphi_2)$$

defined by composition of morphisms in $\underline{C}$.

We shall suppose that there exists a functor

$$\text{Der} : \underline{\text{Mor}}\ \underline{c} \to \underline{\text{Ab}}$$

with properties:

(Der 1) There exists a map:

$$\mu : \pi^{-1}(\varphi) \times \text{Der}(\varphi) \to \pi^{-1}(\varphi)$$

and a partially defined map

$$\nu : \pi^{-1}(\varphi) \times \pi^{-1}(\varphi) \to \text{Der}(\varphi)$$

defined on the subset of those pairs (λ_1,λ_2) having same "source" and same "aim". These maps should satisfy the following relations

$$\mu(\lambda,\alpha+\beta) = \mu(\mu(\lambda,\alpha),\beta)$$

$\nu(\lambda_1,\lambda_2) = \alpha$ is equivalent to $\lambda_1 = \mu(\lambda_2,\alpha)$.

(i.e. the subsets of $\pi^{-1}(\varphi)$ consisting of all morphisms with fixed source and aim are principal homogeneous spaces over $\text{Der}(\varphi)$.)

(Der 2) Suppose φ_1 and φ_2 can be composed in c, then the diagram

$$\begin{array}{ccc}
\pi^{-1}(\varphi_1)\times\pi^{-1}(\varphi_2) & \xrightarrow{\ m\ } & \pi^{-1}(\varphi_1\circ\varphi_2) \\
\uparrow{\scriptstyle \mu\times\mu} & & \nwarrow{\scriptstyle \mu} \\
(\pi^{-1}(\varphi_1)\times \mathrm{Der}(\varphi_1))\times(\pi^{-1}(\varphi_2)\times \mathrm{Der}(\varphi_2)) & \xrightarrow[\delta]{} & \pi^{-1}(\varphi_1\circ\varphi_2)\times \mathrm{Der}(\varphi_1\circ\varphi_2)
\end{array}$$

commutes, with δ defined by:

$$\delta((\lambda_1,\alpha),(\lambda_2,\beta)) = (m(\lambda_1,\lambda_2),\ \mathrm{Der}(id,\varphi_2)(\alpha) + \mathrm{Der}(\varphi_1,id)(\beta))$$

Note that $(id,\varphi_2):\varphi_1 \to \varphi_1\circ\varphi_2$ and $(\varphi_1,id):\varphi_2\to\varphi_1\circ\varphi_2$ are morphisms in Mor $\underline{c}$, since the diagrams

$$\begin{array}{ccc}
* & \xrightarrow{\ \varphi_1\ } & * \\
{\scriptstyle \varphi_1\circ\varphi_2}\downarrow & & \downarrow{\scriptstyle \varphi_2} \\
* & \xleftarrow[\ 1\ ]{} & *
\end{array}
\qquad\qquad
\begin{array}{ccc}
* & \xrightarrow{\ 1\ } & * \\
{\scriptstyle \varphi_1\circ\varphi_2}\downarrow & & \downarrow{\scriptstyle \varphi_1} \\
* & \xleftarrow[\ \varphi_2\ ]{} & *
\end{array}$$

commute.

We shall from now on use the following notations:

$$\begin{aligned}
\varphi_1\beta &= \mathrm{Der}(\varphi_1,id)(\beta)\\
\alpha\varphi_2 &= \mathrm{Der}(id,\varphi_2)(\alpha)\\
\lambda_1-\lambda_2 &= \nu(\lambda_1,\lambda_2)\\
\lambda_1+\alpha &= \mu(\lambda_1,\alpha)
\end{aligned}$$

A functor Der with these properties will be called a derivation functor associated to π.

There are some obvious examples.

Ex.1. Let $\pi : R \to S$ be a surjective homomorphism of rings. Let $I = \ker\pi$ and suppose $I^2 = 0$. Consider the category $\underline{C}$ of flat R-algebras and the category $\underline{c}$ of flat S-algebras. Tensorization with S over R defines a functor

$$\pi : \underline{C} \to \underline{c}$$

and the ordinary derivation functor

$$\mathrm{Der} : \underline{\mathrm{Mor}}\ \underline{c} \to \underline{\mathrm{Ab}}$$

given by:

$$\mathrm{Der}(\varphi) = \mathrm{Der}_S(A, B \underset{S}{\otimes} I)$$

where $\varphi : A \to B$ defines the A-module structure on $B \underset{S}{\otimes} I$, is a <u>derivation functor for</u> π.

Ex.2. Let $\underline{C}_o$ be the full subcategory of $\underline{C}$ defined by the free R-algebras (i.e. the polynomial rings over R in any set of variables), and let $\underline{c}_0$ be the full subcategory of $\underline{c}$ defined by the free S-algebras. As above the ordinary derivation functor induces a derivation functor for the restriction π_0 of π to $\underline{C}_o$.

Ex.3. Let $\pi : R \to S$ be as before and let $\underline{C}$ be the category of R-flat affine group schemes over R and $\underline{c}$ the category of S-flat affine group schemes over S.

Tensorization by S over R defines a functor

$$\pi : \underline{C} \to \underline{c}$$

Let φ be an object in $\underline{\mathrm{Mor}}\ \underline{c}$ (i.e. $\varphi : \mathrm{Spec}(B) \to \mathrm{Spec}(A)$ is a homomorphism of S-flat affine group schemes over S) and consider

$$\mathrm{Der}(\varphi) = \{\xi \in \mathrm{Der}_S(A, B \otimes \ker\pi) \mid \xi \circ \mu_B = \mu_A \circ (\varphi \otimes \xi + \xi \otimes \varphi)\}$$

where $\mu_A : A \to A \otimes A$ and $\mu_B : B \to B \otimes B$ are the comultiplications defining the group scheme structure on Spec(A) and Spec(B) respectively.

Then Der is a <u>derivation functor for</u> π.

Remark. If $\pi^{-1}(\varphi)$ is empty then the conditions (Der 1) and (Der 2) are vacuous.

(1.2) *Obstructions for the existence of sections of functors.*

Consider a diagram of functors of small categories

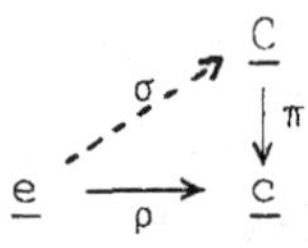

Assume π has a derivation functor $\mathrm{Der} : \mathrm{Mor}\ \underline{c} \to \underline{\mathrm{Ab}}$.
Given ρ and π when does there exist a *section* i.e. a functor σ such that $\sigma \circ \pi = \rho$?
We must certainly require $\pi^{-1}(\rho(\varphi)) \neq \emptyset$ for all $\varphi \in \underline{\mathrm{Mor}}\ \underline{e}$, and, moreover, there should exist a *quasisection* i.e. a *set theoretical* map $\sigma' : \mathrm{Mor}\ \underline{e} \to \mathrm{Mor}\ \underline{C}$ such that $\sigma' \circ \pi = \rho$ and such that $\sigma'(\varphi_1)$ and $\sigma'(\varphi_2)$ may be composed if φ_1 and φ_2 are composable, the source and aim of $\sigma'(\varphi_1) \circ \sigma'(\varphi_2)$ and $\sigma'(\varphi_1 \circ \varphi_2)$ being the same.
Given such a quasisection σ' we deduce a map $\sigma_0 : \mathrm{ob}\,\underline{e} \to \mathrm{ob}\,\underline{C}$ which we shall call the *stem* of the quasisection σ'.

Definition (1.2.1) Given any functor

$$F : \mathrm{Mor}\ \underline{e} \to \underline{\mathrm{Ab}}$$

we shall denote by $H^i(\underline{e},F)$ the groups $\varprojlim_{\mathrm{Mor}\,\underline{e}}^{(i)} F$. We shall refer to $H^i(\underline{e},F)$ as the cohomology of $\underline{e}$ with values in F.

Theorem (1.2.2) In the situation above, suppose given a quasisection σ' of π. Then there exists an obstruction $o(\sigma') = o(\sigma_0) \in H^2(\underline{e},\mathrm{Der})$ such that $o(\sigma_0) = 0$ if and only if there exists a section σ of π with the same stem σ_0 as σ'. Moreover, if $o(\sigma') = 0$ then the set of sections having the stem σ_0, modulo isomorphisms reducing to the identity, is a principal homogeneous space over

$$H^1(\underline{e},\mathrm{Der}).$$

Proof. Consider the complex $D^\cdot = D^\cdot(\mathrm{Der})$ of abelian groups defined by

$$D^0(\mathrm{Der}) = \prod_{e \in \mathrm{ob}\,\underline{e}} \mathrm{Der}(1_e)$$

$$D^n(\mathrm{Der}) = \prod_{e_0 \xrightarrow{\psi_1} e_1 \to \cdots \to e_{n-1} \xrightarrow{\psi_n} e_n} \mathrm{Der}(\psi_1 \circ \psi_2 \circ \cdots \circ \psi_n) \qquad n \geq 1$$

where the indices are chains of morphisms in $\underline{e}$, and where

$$d^n : D^n \to D^{n+1}$$

is defined by:

$$(d^0\xi)(\psi_1) = \psi_1\, \xi_{e_1} - \xi_{e_0}\, \psi_1$$

$$(d^n\xi)(\psi_1,\dots,\psi_{n+1}) = \psi_1 \xi(\psi_2,\dots,\psi_{n+1}) +$$

$$\sum_{i=1}^{n} (-1)^i \xi(\psi_1,\dots,\psi_i \circ \psi_{i+1},\dots,\psi_{n+1}) + (-1)^{n+1}\xi(\psi_1,\dots,\psi_n)\psi_{n+1}$$

for $n \geq 1$.

One easily verifies that $d^n \circ d^{n+1} = 0$ for all $n \geq 0$.

Lemma (1.2.3) $H^n(D^\cdot) \simeq \varprojlim^{(n)}_{\underline{\mathrm{Mor}}\ \underline{e}} \mathrm{Der}$

The proof will be given in (1.3).

Now consider the quasisection σ' and define the element $O(\sigma')$ of D^2 by:

$$O(\sigma')(\psi_1,\psi_2) = \sigma'(\psi_1 \circ \psi_2) - \sigma'(\psi_1) \circ \sigma'(\psi_2) .$$

By assumption $O(\sigma')(\psi_1,\psi_2) \in \mathrm{Der}(\psi_1 \circ \psi_2)$.

In fact $O(\sigma') \in \ker d^2$ since

$$(d^2O(\sigma'))(\psi_1,\psi_2,\psi_3) = \psi_1 O(\sigma')(\psi_2,\psi_3) - O(\sigma')(\psi_1 \circ \psi_2,\psi_3)$$
$$+ O(\sigma')(\psi_1,\psi_2 \circ \psi_3) - O(\sigma')(\psi_1,\psi_2)\psi_3$$
$$= \psi_1(\sigma'(\psi_2 \circ \psi_3) - \sigma'(\psi_2) \circ \sigma'(\psi_3)) - (\sigma'(\psi_1 \circ \psi_2 \circ \psi_3) - \sigma'(\psi_1 \circ \psi_2) \circ \sigma'(\psi_3))$$
$$+ (\sigma'(\psi_1 \circ \psi_2 \circ \psi_3) - \sigma'(\psi_1) \circ \sigma'(\psi_2 \circ \psi_3)) - (\sigma'(\psi_1 \circ \psi_2) - \sigma'(\psi_1) \circ \sigma'(\psi_2))\psi_3$$
$$= (\sigma'(\psi_1) \circ \sigma'(\psi_2 \circ \psi_3) - \sigma'(\psi_1) \circ \sigma'(\psi_2) \circ \sigma'(\psi_3))$$
$$- (\sigma'(\psi_1 \circ \psi_2 \circ \psi_3) - \sigma'(\psi_1 \circ \psi_2)\sigma'(\psi_3))$$
$$+ (\sigma'(\psi_1 \circ \psi_2 \circ \psi_3) - \sigma'(\psi_1) \circ \sigma'(\psi_2 \circ \psi_3))$$
$$- (\sigma'(\psi_1 \circ \psi_2)\sigma'(\psi_3) - \sigma'(\psi_1) \circ \sigma'(\psi_2) \circ \sigma'(\psi_3))$$
$$= 0 .$$

It follows that $O(\sigma')$ defines an element $o(\sigma') \in H^2(D^\cdot)$.

Suppose $o(\sigma') = 0$, then there is a $\xi \in D^1$ such that $d\xi = O(\sigma')$.

Now put

$$\sigma(\varphi) = \sigma'(\varphi) + \xi(\varphi)$$

Then $\sigma(\psi_1 \circ \psi_2) - \sigma(\psi_1) \circ \sigma(\psi_2)$

$$= (\sigma'(\psi_1 \circ \psi_2) + \xi(\psi_1 \circ \psi_2)) - (\sigma'(\psi_1) + \xi(\psi_1)) \circ (\sigma'(\psi_2) + \xi(\psi_2))$$
$$= \sigma'(\psi_1 \circ \psi_2) - \sigma'(\psi_1) \circ \sigma'(\psi_2) - \quad (\psi_1 \xi(\psi_2) - \xi(\psi_1 \circ \psi_2)$$
$$+ \xi(\psi_1)\psi_2) \quad = O(\sigma')_{(\psi_1,\psi_2)} - (d\xi)_{(\psi_1,\psi_2)} = 0 .$$

i.e. σ is a functor, (we easily find that $\sigma(1_e) = 1_{\sigma_0(e)}$).

Obviously the stem of σ is equal to the stem of σ' (i.e. $= \sigma_0$).

Now let σ_1 and σ_2 be two sections of π with the same stem

σ_o. Then $(\sigma_1 - \sigma_2)$ defines an element in D^1, given by:

$$(\sigma_1 - \sigma_2)(\psi) = \sigma_1(\psi) - \sigma_2(\psi) .$$

Since σ_1 and σ_2 both are sections $(d^1(\sigma_1-\sigma_2))(\psi_1,\psi_2)$ $= \psi_1(\sigma_1-\sigma_2)(\psi_2) - (\sigma_1-\sigma_2)(\psi_1 o \psi_2) + (\sigma_1-\sigma_2)(\psi_1)\psi_2 = 0$, and therefore $(\sigma_1-\sigma_2)$ defines an element in $H^1(D^\cdot)$.

Suppose this element is zero, then there exists an element $\xi \in D^o$ such that

$$\sigma_1(\psi) - \sigma_2(\psi) = \psi\xi - \xi\psi$$

i.e.

$$\sigma_1(\psi) \circ (1_{\sigma_o(e_0)} - \xi e_1) = (1_{\sigma_o(e_0)} - \xi e_0) \circ \sigma_2(\psi)$$

for all morphisms $\psi: e_0 \to e_1$ of $\underline{e}$.

Conversely, suppose $s \in H^1(D^\cdot)$ is represented by $\xi \in D^1$ then given any section σ of π, $\xi + \sigma$ is another section with the same stem as σ.

QED.

(1.3) Resolving functors for $\varprojlim$.

Let $\underline{c}$ be any small category and denote by $\underline{Ab}^{\underline{c}^o}$ the category of abelian functors on $\underline{c}^o$. Recall (see (La 1)) the standard resolving complex

$$C^\cdot : \underline{Ab}^{\underline{c}^o} \to \underline{\text{Compl.ab.gr}}.$$

defined by

$$C^p(G) = \prod_{c_0 \xrightarrow{\psi_1} c_1 \xrightarrow{\psi_2} \cdots \xrightarrow{\psi_p} c_p} G(c_0)$$

with differential $d^p : C^p(G) \to C^{p+1}(G)$ given by:

$$(d^p(\xi))(\psi_1,\cdots,\psi_{p+1}) = G(\psi_1)(\xi(\psi_2,\cdots,\psi_{p+1}))$$
$$+ \sum_{i=1}^{n} (-1)^i \xi(\psi_1,\cdots,\psi_i \circ \psi_{i+1},\cdots,\psi_{p+1}) + (-1)^{n+1}\xi(\psi_1,\cdots,\psi_p) .$$

The basic properties of $C^\cdot = C^\cdot(\underline{c}^o,-)$ are the following, see the Appendix,

1) $C^\cdot(\underline{c}^o,-)$ is an exact functor

2) $H^n(C^\cdot(\underline{c}^o,-)) = \varprojlim_{\underline{c}^o}{}^{(n)}$ for $n \geq 0$.

Now let F be any abelian functor on $\mathrm{Mor}\,\underline{c}$ (i.e. F is an object of $\underline{Ab}^{\mathrm{Mor}\,\underline{c}}$) and put

$$D^p(F) = D^p(\underline{c},F) = \prod_{c_0 \xrightarrow[\psi_1]{} c_1 \to \cdots \xrightarrow[\psi_p]{} c_p} F(\psi_1 \circ \cdots \circ \psi_p) .$$

Let d^p be the homomorphism $D^p(F) \to D^{p+1}(F)$ defined by

$$(d^p(\xi))(\psi_1,\ldots,\psi_{p+1}) = F(\psi_1,1_{c_{p+1}})(\xi(\psi_2,\ldots,\psi_{p+1}))$$

$$+ \sum_{i=1}^{p} (-1)^i \xi(\psi_1,\ldots,\psi_i \circ \psi_{i+1},\ldots,\psi_{p+1}) + (-1)^{p+1} F(1_{c_0},\psi_{p+1})(\xi(\psi_1,\ldots,\psi_p))$$

(Remember that $(\psi_1,1_{c_{p+1}})$ is a morphism

$$\psi_2 \circ \cdots \circ \psi_{p+1} \to \psi_1 \circ \cdots \circ \psi_{p+1}$$

in $\mathrm{Mor}\,\underline{c}$ and that $(1_{c_0},\psi_{p+1})$ is a morphism

$$\psi_1 \circ \cdots \circ \psi_p \to \psi_1 \circ \cdots \circ \psi_{p+1}$$

in $\mathrm{Mor}\,\underline{c}$).

It is easy to check that $(D^p(F),d^p)_{p\geq 0}$ is a complex of abelian groups defining a functor

$$D^\cdot : \underline{Ab}^{\mathrm{Mor}\,\underline{c}} \to \underline{\mathrm{Compl}}.\underline{\mathrm{ab}}.\underline{\mathrm{gr}}. .$$

<u>Lemma (1.3.1)</u> The functor $D^\cdot = D^\cdot(\underline{c},-)$ has the following properties:

1) $D^\cdot(\underline{c},-)$ is exact

2) $H^n(D^\cdot(\underline{c},-)) = \varprojlim_{\mathrm{Mor}\,\underline{c}}{}^{(n)}$ for $n \geq 0$.

Proof. Let L be the constant functor on $\underline{\text{Mor}}\ \underline{c}$, i.e. $L(\varphi) = \mathbb{Z}$ for all φ.

We shall construct a projective resolution of L in $\text{Ab}^{\underline{\text{Mor}\ c}}$.

Let $\varphi : x \to y$ be any object of $\underline{\text{Mor}}\ \underline{c}$ and consider the sets

$$\Delta^0(\varphi) = \{x \overset{\varepsilon}{\to} c_0 \overset{\rho}{\to} y \mid \varepsilon \circ \rho = \varphi\},$$

$$\Delta^n(\varphi) = \{x \overset{\varepsilon}{\to} c_0 \overset{\psi_1}{\to} c_1 \to \cdots \to c_{n-1} \overset{\psi_n}{\to} c_n \overset{\rho}{\to} y \mid \varepsilon \circ \psi_1 \circ \cdots \circ \psi_n \circ \rho = \varphi\}.$$

There exist maps:

$$\eta_i^n : \Delta^n(\varphi) \to \Delta^{n+1}(\varphi)$$

$$\delta_i^n : \Delta^n(\varphi) \to \Delta^{n-1}(\varphi)$$

defined by:

$$\eta_i^n(x \overset{\varepsilon}{\to} c_0 \overset{\psi_1}{\to} c_1 \to \cdots \to c_{n-1} \overset{\psi_n}{\to} c_n \overset{\rho}{\to} y) = (x \overset{\varepsilon}{\to} c_0 \to \cdots \to c_i \overset{id}{\to} c_i \to \cdots \to c_n \overset{\rho}{\to} y)$$

$$\delta_i^n(x \overset{\varepsilon}{\to} c_0 \overset{\psi_1}{\to} c_1 \to \cdots \to c_{n-1} \overset{\psi_n}{\to} c_n \overset{\rho}{\to} y) = \begin{cases} (x \overset{\varepsilon\circ\psi_1}{\to} c_1 \overset{\psi_2}{\to} c_2 \to \cdots \to c_n \overset{\rho}{\to} y), & i = 0 \\ (x \overset{\varepsilon}{\to} c_0 \to \cdots \to c_{i-1} \overset{\psi_i\circ\psi_{i+1}}{\to} c_{i+1} \to \cdots \to c_n \overset{\rho}{\to} y), & 0<i<n \\ (x \overset{\varepsilon}{\to} c_0 \to \cdots \to c_{n-1} \overset{\psi_n\circ\rho}{\to} y), & i = n \end{cases}$$

giving $\Delta^n(\varphi)$, $n \geq 0$ the structure of a simplicial set.

Moreover for each $n \geq 0$, $\Delta^n(\varphi)$ is functorial in φ defining a functor

$$\Delta : \underline{\text{Mor}}\ \underline{c} \to \underline{\text{Simplicial}}\ \underline{\text{sets}}$$

Composing Δ with the functor $C.(-,\mathbb{Z})$ we have constructed a complex of functors

$$C. : \underline{\text{Mor}}\ \underline{c} \to \underline{\text{Ab}}$$

Now, by a standard argument we construct a contracting homotopy for $C.$ thereby proving

$$H_i(C.) = \begin{cases} L & \text{for } i = 0 \\ 0 & \text{for } i \neq 0 \end{cases}$$

Moreover

$$C_n(\varphi) = \coprod_{\substack{(\varepsilon,\rho):\varphi'\to\varphi \\ \text{in } \underline{\text{Mor } c}}} \{ \coprod_{\substack{\psi_1,\ldots,\psi_n \\ \psi_1\circ\ldots\circ\psi_n = \varphi'}} \mathbb{Z} \}$$

Consequently, see the Appendix, it follows that each C_n is projective as an object of $\underline{\text{Ab}}^{\underline{\text{Mor } c}}$.

Therefore $C.$ is a projective resolution of L in $\underline{\text{Ab}}^{\underline{\text{Mor } c}}$.

Since

$$\text{Mor}_{\underline{\text{Ab}}^{\underline{\text{Mor } c}}}(C_n, F) = \prod_{c_0 \xrightarrow{\psi_1} c_1 \to \cdots \to c_{n-1} \xrightarrow{\psi_n} c_n} F(\psi_1\circ\cdots\circ\psi_n)$$

we find by a dull computation that

$$D^\bullet(F) = \text{Mor}_{\underline{\text{Ab}}^{\underline{\text{Mor } c}}}(C., F)$$

thereby proving the lemma.

QED.

Chapter 2

Lifting algebras and morphisms of algebras.

In this chapter we establish the basic obstruction calculus for lifting commutative algebras. Given a diagram of morphisms of commutative rings,

$$(*) \qquad \begin{array}{ccc} R & & \\ \pi\downarrow & & \\ S & \to & A \end{array}$$

when will there exist a completed diagram

$$(**) \qquad \begin{array}{ccc} R & \to & A' \\ \pi\downarrow & & \downarrow\pi' \\ S & \to & A \end{array}$$

with $A = A' \otimes_R S$, $\mathrm{Tor}_1^R(A',S) = 0$?

An R-algebra A' with this property is called a lifting of the S-algebra A to R.

We shall assume π is surjective and that $(\ker\pi)^2 = 0$. Using the results of Chapter 1. we shall show that $(*)$ defines an obstruction $o \in H^2(S,A;A \otimes_S \ker\pi)$ s.t. $o = 0$ iff there exists a lifting $(**)$.

Moreover the set of equivalence classes of such liftings is a principal homogenous fiberspace over $H^1(S,A;A \otimes_S \ker\pi)$.

Corresponding results for morphisms of S-algebras and compositions of such follow easily.

(2.1) Leray spectral sequence for $\varprojlim$.

Let $\underline{c}$ be any small category and let c be an object of $\underline{c}$. Consider the contravariant functor $C(\mathbb{Z},c)$ defined by:

$$C(\mathbb{Z},c)(c') = \coprod_{c' \underset{\varphi}{\to} c} \mathbb{Z}$$

We may easily show that these functors are projective objects in $\underline{Ab}^{\underline{c}^o}$, see the Appendix.

Suppose $\underline{M}$ is a full subcategory of $\underline{c}$ and consider the restriction of $C(\mathbb{Z},c)$ to $\underline{M}$. Let F be any contravariant functor on $\underline{M}$ with values in $\underline{Ab}$, then we find,

$$\underline{Ab}^{\underline{M}^o}(C(\mathbb{Z},c),F) \simeq \varprojlim_{(\underline{M}/c)^o} F .$$

Now, suppose $c_0 \underset{\varphi}{\to} c$ in $\underline{c}$ is an $\underline{M}$-epimorphism, i.e. $c_0 \in \text{ob}\,\underline{M}$ and the map

$$\text{Mor}(c',c_0) \to \text{Mor}(c',c)$$

is surjective for every $c' \in \text{ob}\,\underline{M}$.

Suppose further that $\underline{c}$ has fibered products and consider the system of morphisms

$$c \leftarrow c_0 \leftleftarrows c_0 \underset{c}{\times} c_0 \overset{\leftarrow}{\underset{\leftarrow}{\leftarrow}} \cdots \vdots\; c_0 \underset{c}{\times} \cdots \underset{c}{\times} c_0 \vdots$$

Put $c_p = \underbrace{c_0 \underset{c}{\times} \cdots \underset{c}{\times} c_0}_{p+1}$ and denote by

$$d_p^i : c_p \to c_{p-1} \qquad i = 0,\dots,p$$

the $p+1$ projection morphisms.

Consider for each d_p^i the corresponding morphism

$\partial_p^i : C(\mathbb{Z},c_p) \to C(\mathbb{Z},c_{p-1})$ and let $\partial_p = \sum_{i=0}^{p} (-1)^i \partial_p^i$. Then $\partial_p \partial_{p-1} = 0$ for all $p \geq 1$.

Lemma (2.1.1) The complex $C. = \{C(\mathbb{Z},c_p),\partial_p\}_{p\geq 0}$ is a resolution of $C(\mathbb{Z},c)$ in $\underline{Ab}^{\underline{M}^o}$.

Proof. See f.ex. [Ar] p. 18.

QED.

Let $F^{\cdot}$ be an injective resolution of F in $\underline{Ab}^{\underline{M}^o}$ and consider the double complex

$$\mathrm{Mor}(C.,F^{\cdot})$$

We shall compute the two associated spectral sequences. But first we have to establish the following lemma.

Lemma (2.1.2) Let $f : \underline{M}/c \to \underline{M}$ be the canonical forgetful functor and let F be injective in $\underline{Ab}^{\underline{M}^o}$, then the composed functor $f \circ F : (\underline{M}/c)^o \to \underline{Ab}$ is injective as an object of $\underline{Ab}^{(\underline{M}/c)^o}$.

Proof. The functor f induces a functor

$$f_* : \underline{Ab}^{\underline{M}^o} \to \underline{Ab}^{(\underline{M}/c)^o}$$

We want to prove that f_* takes injectives into injectives. To prove this we construct a left adjoint

$$\rho : \underline{Ab}^{(\underline{M}/c)^o} \to \underline{AB}^{\underline{M}^o}$$

Let G be an object of $\underline{Ab}^{(\underline{M}/c)^o}$ and put

$$\rho(G)(m) = \coprod_{\varphi \in \mathrm{Mor}(m,c)} G(m \xrightarrow{\varphi} c)$$

so that $\rho(G)$ is an object of $\underline{Ab}^{\underline{M}^o}$.

One easily checks that there is a canonical isomorphism

$$\mathrm{Mor}(\rho(G),F) \simeq \mathrm{Mor}(G,f_*(F))$$

proving that ρ is left adjoint to f_*. Since ρ is exact we know that f_* takes injectives into injectives. QED.

Going back to the double complex $\mathrm{Mor}(C.,F^{\cdot})$ we find the E_2 terms of the two associated spectral sequences:

$$'E_2^{p,q} = H^p(H^q(\widehat{\mathrm{Mor}(C.,F^{\cdot})}))$$

$$''E_2^{p,q} = H^p(\mathrm{Mor}(H_q(C.),F^{\cdot}))$$

We know already that

$$''E_2^{p,q} = 0 \quad \text{for} \quad q \neq 0$$

$$''E_2^{n,0} = H^n(\varprojlim_{(\underline{M}/c)^o} (F^{\cdot}))$$

and by Lemma (2.1.2) we deduce

$$''E^{n,0} = \varprojlim_{(\underline{M}/c)^o} {}^{(n)}F .$$

Since

$$\mathrm{Mor}(C_p,F^{\cdot}) = \varprojlim_{\underline{M}/c_p} F^{\cdot} ,$$

we find, using Lemma (2.1.1) once more, that

$$'E_2^{p,q} = H^p(\varprojlim_{\underline{M}/c.} {}^{(q)}F).$$

We have proved the following theorem.

<u>Theorem (2.1.3)</u> Let $\underline{M} \subseteq \underline{c}$ and $\varphi : c_0 \to c$ be given as above. Then there exists a <u>Leray spectral sequence</u> given by:

$$E_2^{p,q} = E_2^{p,q}(\underline{M}) = H^p(\varprojlim_{(\underline{M}/c.)^o} {}^{(q)}F)$$

converging to

$$\varprojlim_{(\underline{M}/c)^{\circ}} {}^{(\cdot)} F .$$

Remark 1. The spectral sequence above is nothing but the Leray spectral sequence associated to the "covering" $\varphi : c_0 \to c$ in an appropriate Grothendieck topology.

2. Since $c_0 \in \text{ob}\, \underline{M}$ the category $\underline{M}/c_0$ has a final object. Therefore $E_2^{0,q} = 0$ for all $q \geq 1$.

We deduce from this the formulas

$$\varprojlim_{(\underline{M}/c)^{\circ}} F \simeq E^{0,0}$$

$$\varprojlim_{(\underline{M}/c)^{\circ}} {}^{(1)} F \simeq E^{1,0} ,$$

and the exact sequence

$$0 \to E_2^{2,0} \to \varprojlim_{(\underline{M}/c)^{\circ}} {}^{(2)} F \to E_2^{1,1} \to E_2^{3,0} \to \varprojlim_{(\underline{M}/c)^{\circ}} {}^{(3)} F .$$

Corollary (2.1.4) Suppose that $\varprojlim_{(\underline{M}/c_j)^{\circ}} {}^{(i)} F = 0$ for $i \geq 1$, $i+j = p$ and for $i+j = p-1$. Then

$$\varprojlim_{(\underline{M}/c)^{\circ}} {}^{(p)} F \simeq E_2^{p,o} .$$

Assume for a moment that there exists a functor $i : \underline{c} \to \underline{Ab}$ commuting with fibered products.

Corollary (2.1.5) Put $g = f \circ i$ and suppose

$$\varinjlim_{\underline{M}/c_p} g = i(c_p) \qquad \text{for all } p \geq 0 .$$

Then

$$\varinjlim^{(1)}_{\underline{M}/c} g = 0 .$$

Proof. Let E be an injective abelian group and consider the functor

$$F(-) = \underline{Ab}(g(-),E) .$$

We know that

$$\underline{Ab}(\varinjlim^{(1)}_{\underline{M}/c} g, E) \simeq \varprojlim^{(1)}_{(\underline{M}/c)^o} F$$

$$= \ker\{\varprojlim_{(\underline{M}/c_1)^o} F \to \varprojlim_{(\underline{M}/c_2)^o} F\} / \mathrm{im}\{\varprojlim_{(\underline{M}/c_0)^o} F \to \varprojlim_{(\underline{M}/c_1)^o} F\}$$

$$= \underline{Ab}(\ker\{i(c_1) \to i(c_0)\}/\mathrm{im}\{i(c_2) \to i(c_1)\},E)$$

But since $i(c_p) = \underbrace{i(c_o) \underset{i(c)}{\times} \cdots \underset{i(c)}{\times} i(c_o)}_{p+1}$

this last group is zero.

Since this holds for all injective abelian groups E we have proved that $\varinjlim^{(1)}_{\underline{M}/c} g = 0$. QED.

Corollary (2.1.6) Let $\underline{M}_o \subseteq \underline{M}$ be two full subcategories of $\underline{c}$. Suppose $\underline{c}$ has fibered products and let $c \in \mathrm{ob}\, \underline{c}$. Assume that $(c, \underline{M}_o, \underline{M})$ satisfies the following conditions:

(c_1) There exists an object c_0 of $\underline{M}_o$ and an $\underline{M}$-epimorphism $\varphi : c_0 \to c$.

(c_2) For any $\underline{M}$-epimorphism $\psi : d_0 \to d$ in $\underline{c}$ with $d_0 \in \underline{M}_0$ there exist objects $e_p \in \underline{M}_0$ and $\underline{M}$-epimorphisms

$$\psi_p : e_p \to \underbrace{d_0 \times_d \dots \times_d d_0}_{p+1} \qquad p \geq 1 .$$

Then we may conclude

$$\lim_{\substack{\leftarrow \\ (\underline{M}/c)^o}}{}^{(\cdot)} \simeq \lim_{\substack{\leftarrow \\ (\underline{M}_0/c)^o}}{}^{(\cdot)}$$

Proof. We first observe that (c_1) and (c_2) together with (2.1.1) imply that there are canonical isomorphisms

(1) $$\lim_{\substack{\leftarrow \\ (\underline{M}/c_p)^o}} \simeq \lim_{\substack{\leftarrow \\ (\underline{M}_0/c_p)^o}}$$

where $c_p = \underbrace{c_0 \times_c \dots \times_c c_0}_{p+1}$.

Now the canonical morphism

$$t^n : \lim_{\substack{\leftarrow \\ (\underline{M}/c)^o}}{}^{(n)} \to \lim_{\substack{\leftarrow \\ (\underline{M}_0/c)^o}}{}^{(n)}$$

induces morphisms of the Leray spectral sequences

$$t_2^{p,q} : E_2^{p,q}(\underline{M}) \to E_2^{p,q}(\underline{M}_0)$$

Using (1) we find isomorphisms

$$t_2^{p,o} : E_2^{p,o}(\underline{M}) \xrightarrow{\simeq} E_2^{p,o}(\underline{M}_0) \qquad p \geq 0 .$$

Thereby proving that t^1 is an isomorphism. By an easy induction argument we may assume that $t_2^{p,q}$ are isomorphisms for all p,q with $p+q \leq n$ or $q < n$. This implies that

$$t_\infty^{p,q} : E_\infty^{p,q}(\underline{M}) \to E_\infty^{p,q}(\underline{M}_0)$$

are isomorphisms for all p,q with $p+q = n$, thereby proving that t^n is an isomorphism. QED.

Corollary (2.1.7) Suppose we have given a commutative diagram of functors of small categories

$$\begin{array}{ccc} \underline{c} & \overset{f}{\to} & \underline{d} \\ \cup| & & \cup| \\ \underline{M} & \underset{f_o}{\to} & \underline{N} \end{array}$$

Suppose $\underline{c}$ and $\underline{d}$ have, and f preserves, finite fibered powers.

Suppose finally that for every object c of $\underline{c}$ there is an $\underline{M}$-epimorphism $m \to c$ with $m \in \text{ob}\,\underline{M}$, and that f transforms $\underline{M}$-epimorphisms into $\underline{N}$-epimorphisms.

Then for every functor $G : \underline{N}^o \to \underline{Ab}$ satisfying the condition

$$\varprojlim_{(\underline{M}/c)^o} f_o \circ G \simeq \varprojlim_{(\underline{N}/f(c))^o} G \qquad \forall c \in \text{ob}\,\underline{c}$$

we have isomorphisms

$$\varprojlim_{(\underline{M}/c)^o}{}^{(i)} f_o \circ G \simeq \varprojlim_{(\underline{N}/f(c)^o}{}^{(i)} G \qquad \forall c \in \text{ob}\,\underline{c}$$

Proof. Consider the Leray spectral sequence $E_2^{p,q}(f_o \circ G)$ (resp. $E_2^{p,q}(G)$) associated to the $\underline{M}$ (resp. $\underline{N}$)-epimorphism $m_o \to c$ (resp. $n_o = f_o(m_o) \to f(c) = d$).

Since f preserves finite fibered powers we find

$$f(m_p) = f(m \times_c \ldots \times_c m) = n_o \times_d \ldots \times_d n_o = n_p .$$

f induces a morphism of the E_2-terms of the spectral sequence

$$f^* : E_2^{p,q}(G) \to E_2^{p,q}(f_o \circ G) .$$

By assumption this is an isomorphism for $q = 0$, $p \geq 0$.

Since $m_o \in \underline{M}$ and $n_o \in \underline{N}$ thus $E_2^{o,q}(G) = E_2^{o,q}(f_o \circ G) = 0$

for $q \geq 1$, f induces isomorphisms

$$E_r^{p,q}(G) \to E_r^{p,q}(f_o \circ G)$$

for all $p,q \geq 0$ with $p+q \leq 1$ and all $r \geq 2$.
Consequently we find:

$$\varprojlim_{(\underline{N}/d)^o}{}^{(1)} G \simeq \varprojlim_{(\underline{M}/c)^o}{}^{(1)} f_o \circ G .$$

c being arbitrary this implies that

$$f^* : E_2^{p,1}(G) \to E_2^{p,1}(f_o \circ G)$$

is an isomorphism for all $p \geq 0$. As above we conclude that

$$f^* : E_r^{p,q}(G) \to E_r^{p,q}(f_o \circ G)$$

is an isomorphism for all $p,q \geq 0$ with $p+q \leq 2$ and all $r \geq 2$. Thus

$$\varprojlim_{(\underline{N}/d)^o}{}^{(2)} G \simeq \varprojlim_{(\underline{M}/c)^o}{}^{(2)} (f_o \circ G) .$$

An easy induction argument proves that

$$f^* : E_r^{p,q}(G) \to E_r^{p,q}(f_o \circ G)$$

is an isomorphism for all $p,q \geq 0$ with $p+q \leq n$ and all $r \geq 2$. Consequently we have proved

$$\varprojlim_{(\underline{N}/d)^o}{}^{(n)} G \simeq \varprojlim_{(\underline{M}/c)^o}{}^{(n)} f_o \circ G$$

QED.

(2.2) Lifting of algebras.

Let S be any commutative ring with unit. Let S-alg denote the category of S-algebras and let S-free denote the category of free S-algebras (i.e. the category of polynomial algebras, in any set of variables, over S).

Let A be any object of $\underline{\text{S-alg}}$ and consider the subcategories $\underline{M}_o$ and $\underline{M}$ of $\text{S-alg}/_A$ where $\underline{M} = \text{S-free}/_A$ and $\underline{M}_o = (\text{S-free}/_A)^{epi}$ is the full subcategory of $\underline{M}$ defined by the epimorphisms $F \to A$.

Thus we have $\underline{M}_o \subseteq \underline{M} \subseteq \text{S-alg}/_A$.

We observe that we have isomorphisms of categories:

$$\underline{M}_o \simeq \underline{M}_o/(A \underset{1_A}{\to} A)$$

$$\underline{M} \simeq \underline{M}/(A \underset{1_A}{\to} A)$$

$$\text{S-}\underline{\text{alg}}/_A \simeq (\text{S-}\underline{\text{alg}}/_A)/(A \underset{1_A}{\to} A).$$

Let f (resp. f_o) be the forgetful functor $\underline{M} \to \text{S-alg}$ (resp. $M_o \to \text{S-alg}$). By straight forward verification we find that $\underline{M}_o \subseteq \underline{M} \subseteq \text{S-alg}/_A$ and the object $(A \overset{1_A}{\to} A)$ satisfy the conditions of Corollary (2.1.6). We therefore conclude

<u>Lemma (2.2.1)</u> There are canonical isomorphisms of functors

$$\varprojlim_{(\text{S-}\underline{\text{free}}/_A)^o}{}^{(n)} \simeq \varprojlim_{(\text{S-}\underline{\text{free}}/_A)^{epi,o}}{}^{(n)} \qquad n \geq 0$$

Let $i : \text{S-}\underline{\text{alg}} \to \underline{\text{Ab}}$ be the forgetful functor, then i commutes with fibered products. Thus Corollary (2.1.5) implies

<u>Lemma (2.2.2)</u> Let $g = f\,i$ (resp. $g_o = f_o i$) be the composed functor, then

$$\varinjlim_{\text{S-}\underline{\text{free}}/_A} g = A, \quad \varinjlim_{\text{S-}\underline{\text{free}}/_A}{}_{(1)} g = 0$$

$$\text{(resp.} \quad \varinjlim_{(\text{S-}\underline{\text{free}}/_A)^{epi}} g_o = A, \quad \varinjlim_{(\text{S-}\underline{\text{free}}/_A)^{epi}}{}_{(1)} g_o = 0)$$

<u>Remark</u>. The isomorphism of (2.2.1) is obviously induced by the natural homomorphism of complexes

$$C^{\cdot}({}^{S\text{-}\underline{free}}/A^{o},-) \to C^{\cdot}(({}^{S\text{-}\underline{free}}/A)^{epi,o},-)$$

Now recall (see (An)) that given any A-module M the algebra cohomology $H^{\cdot}(S,A;M)$ is defined by:

$$H^{n}(S,A;M) = \varprojlim_{(S\text{-}\underline{free}/_{A})^{o}}{}^{(n)} \; Der_{S}(-,M)$$

where

$$Der_{S}(-,M) : (S\text{-}\underline{free}/_{A})^{o} \to \underline{Ab}$$

is the functor defined by:

$$Der_{S}(\varphi \overset{F}{\underset{A}{\downarrow}},M) = Der_{S}(F,M)$$

where it is understood that M is considered as an F-module via φ.

Lemma (2.2.1) therefore tells us that we may compute $H^{n}(S,A;M)$ using only the subcategory $(S\text{-}\underline{free}/_{A})^{epi}$ of $(S\text{-}\underline{free}/_{A})$, or stated in a form we shall need later on: the homomorphism of complexes

$$C^{\cdot}((S\text{-}\underline{free}/_{A})^{o},Der_{S}(-,M)) \to C^{\cdot}((S\text{-}\underline{free}/_{A})^{epi,o},Der_{S}(-,M))$$

is a quasiisomorphism (i.e. induces isomorphisms in cohomology).

Consider any S-module I and let

$$Der_{S}(-,-\underset{S}{\otimes} I) : \underline{Mor}(S\text{-}\underline{free}/_{A}) \to \underline{Ab}$$

be the functor defined by:

$$Der_{S}(-,-\otimes I)\begin{pmatrix} F_{0} & \xrightarrow{\alpha_{1}} & F_{1} \\ & {\scriptstyle\delta_{o}}\searrow \; \swarrow{\scriptstyle\delta_{1}} & \\ & A & \end{pmatrix} = Der_{S}(F_{0},F_{1}\underset{S}{\otimes} I)$$

where $F_1 \underset{S}{\otimes} I$ is considered as an F_0-module via the morphism α_1.

Let

$$\mathrm{Der}_S(-,A \underset{S}{\otimes} I) : \underline{\mathrm{Mor}}(S\text{-}\underline{\mathrm{free}}/_A) \to \underline{\mathrm{Ab}}$$

be the functor defined by

$$\mathrm{Der}_S(-,A \underset{S}{\otimes} I)\left(\begin{array}{ccc} F_0 & \xrightarrow{\alpha_1} & F_1 \\ & {}_{\delta_0}\searrow \quad \swarrow_{\delta_1} & \\ & A & \end{array}\right) = \mathrm{Der}_S(F_0, A \underset{S}{\otimes} I)$$

where $A \underset{S}{\otimes} I$ is considered as an F_0-module via the morphism δ_0 $(= \alpha_1 \delta_1)$.

Obviously there is a morphism of functors

$$\mathrm{Der}_S(-,- \underset{S}{\otimes} I) \to \mathrm{Der}_S(-,A \otimes I) .$$

The restriction of this morphism to the subcategory $\underline{\mathrm{Mor}}(S\text{-}\underline{\mathrm{free}}/_A)^{\mathrm{epi}}$ of $\underline{\mathrm{Mor}}(S\text{-}\underline{\mathrm{free}}/_A)$ is moreover surjective.

Notice that by construction

$$D^{\cdot}((S\text{-}\underline{\mathrm{free}}/_A),\mathrm{Der}_S(-,A \underset{S}{\otimes} I)) = C^{\cdot}((S\text{-}\underline{\mathrm{free}}/_A)^{o},\mathrm{Der}_S(-,A \underset{S}{\otimes} I))$$

$$D^{\cdot}((S\text{-}\underline{\mathrm{free}}/_A)^{\mathrm{epi}},\mathrm{Der}_S(-,A \otimes I)) = C^{\cdot}((S\text{-}\underline{\mathrm{free}}/_A)^{\mathrm{epi},o},\mathrm{Der}_S(-,A \underset{S}{\otimes} I))$$

Thus there is a commutative diagram of complexes

$$\begin{array}{ccc} D^{\cdot}((S\text{-}\underline{\mathrm{free}}/_A),\mathrm{Der}_S(-,- \underset{S}{\otimes} I)) & \xrightarrow{l} & C^{\cdot}((S\text{-}\underline{\mathrm{free}}/_A)^{o},\mathrm{Der}_S(-,A \otimes I)) \\ \downarrow k & & i \downarrow \\ D^{\cdot} = D^{\cdot}((S\text{-}\underline{\mathrm{free}}/_A)^{\mathrm{epi}},\mathrm{Der}_S(-,- \underset{S}{\otimes} I)) & \xrightarrow[j]{} & C^{\cdot}((S\text{-}\underline{\mathrm{free}}/_A)^{\mathrm{epi},o},\mathrm{Der}_S(-,A \otimes I)) = C^{\cdot} \end{array}$$

in which i is a quasiisomorphism and j is a surjection.

Put:

$$K^{\cdot} = \ker j .$$

Now let

$$\pi : R \to S$$

be any surjective homomorphism of commutative rings and consider the diagram

$$\underline{e} = \{R \underset{\pi}{\to} S \to A\}$$

<u>Definition (2.2.3)</u> A lifting of $\underline{e}$, or a lifting of A to R , is a commutative diagram of commutative rings

$$\begin{array}{ccc} R & \to & A' \\ \downarrow & & \downarrow \\ S & \to & A \end{array}$$

such that:

(1) $A' \otimes_R S \xrightarrow{\sim} A$

(2) $\mathrm{Tor}_1^R(A', S) = 0$

Abusing the language we shall usually call A' a lifting of A to R .

<u>Definition (2.2.4)</u> Two liftings, A' and A'' , of A to R are equivalent (written $A' \sim A''$), if there exists an isomorphism of rings

$$\theta : A' \to A''$$

such that the following diagram commutes

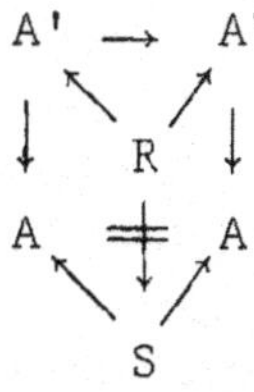

The set of liftings of A to R modulo this equivalence relation is denoted

$$\mathrm{Def}(\underline{e}) = \mathrm{Def}(R \to S \to A)\ .$$

The purposc of this paragraph is to answer the following two questions

1) When does there exist liftings of A to R ?

2) If there do exist some, how many are there?

As usual the answers given will be rather formal and only partial.

In fact we shall have to assume that

$$(\ker\pi)^2 = 0\ ,$$

implying that $I = \ker\pi$ has a natural structure of S-module. Notice that in this case we already know ((1.1) Ex. 2) that the functor

$$\mathrm{Der}_S(-,-\underset{S}{\otimes} I) : \underline{\mathrm{Mor}}(S\text{-}\underline{\mathrm{free}}) \to \underline{\mathrm{Ab}}$$

is a derivation functor for the functor

$$-\underset{R}{\otimes} S : R\text{-}\underline{\mathrm{free}} \to S\text{-}\underline{\mathrm{free}}\ ,$$

the restriction of $-\underset{R}{\otimes} S$ to the subcategory R-free of R-$\underline{\mathrm{Alg}}$.

Suppose there exist a section σ of $-\underset{R}{\otimes} S$ in the diagram,

$$\begin{array}{ccc} & & R\text{-free} \\ & & \downarrow\ -\underset{R}{\otimes} S \\ S\text{-}\underline{\mathrm{free}}/A & \xrightarrow[f]{} & S\text{-}\underline{\mathrm{free}} \end{array}$$

then an easy argument shows that the R-algebra

$$A' = \varinjlim_{S\text{-}\underline{\mathrm{free}}/A} (f\circ\sigma)$$

where f is the forgetfull functor, defines a lifting of A to R .

Now, clearly, the existence of such a section σ of $- \underset{R}{\otimes} S$ is too much to hope for, but the idea, properly modified, is still good.

In fact there are lots of quasisections σ' of $- \underset{R}{\otimes} S : R\text{-}\underline{\text{free}} \to S\text{-}\underline{\text{free}}$ (but only one stem). Picking one we find an obstruction cocycle $O(\sigma')$ in $D^2(S\text{-}\underline{\text{free}}\ , \mathrm{Der}_S(-,- \underset{S}{\otimes} I))$ (see (1.2)). Obviously the forgetfull functor f defines a morphism of complexes

$$D^{\cdot}(S\text{-}\underline{\text{free}},\mathrm{Der}_S(-,- \underset{S}{\otimes} I)) \to D^{\cdot}(S\text{-}\underline{\text{free}}/_A,\mathrm{Der}_S(-,- \underset{S}{\otimes} I)) .$$

Thus $O(\sigma')$ defines a 2-cocycle $O'(\sigma',A)$ of $D^{\cdot}(S\text{-}\underline{\text{free}}/_A),\mathrm{Der}_S(-,- \underset{S}{\otimes} I))$, which maps to a 2-cocycle

$$O(\sigma',A) = l(O'(\sigma'.A)) \in C^2((S\text{-}\underline{\text{free}}/_A)^o,\mathrm{Der}_S(-,A \underset{S}{\otimes} I))$$

under the morphism l (see diagram above).

We already know that the corresponding cohomology class

$$o(\pi,A) \in H^2(S,A;\ A \underset{S}{\otimes} I)$$

does not depend upon the choice of quasisection σ' .

Moreover we shall prove the following

<u>Theorem (2.2.5)</u> There exists an obstruction

$$o(\pi,A) \in H^2(S,A;\ \otimes I)$$

such that $o(\pi,A) = 0$ if and only if there exists a lifting of A to R . In that case $\mathrm{Def}(R \to S \to A)$ is a principal homogeneous space over $H^1(S,A;\ A \underset{S}{\otimes} I)$.

<u>Proof</u>. Consider the diagrams of functors

$$\begin{array}{ccc} & & \text{R-}\underline{\text{free}} \\ & & \downarrow -\underset{R}{\otimes} S = u \\ (\text{S-}\underline{\text{free}}/_A) & \underset{f}{\rightarrow} & \text{S-}\underline{\text{free}} \end{array} \qquad \begin{array}{ccc} & & \text{R-}\underline{\text{free}} \\ & & \downarrow -\underset{R}{\otimes} S = u_o \\ (\text{S-}\underline{\text{free}}/_A)^{epi} & \underset{f_o}{\rightarrow} & \text{S-}\underline{\text{free}} \end{array}$$

<u>Definition (2.2.6)</u> A map

$$\sigma' : \text{mor}(\text{S-}\underline{\text{free}}/_A) \rightarrow \text{mor}(\text{R-}\underline{\text{free}})$$

(resp. σ_o' : $\text{mor}(\text{S-}\underline{\text{free}}/_A)^{epi} \rightarrow \text{mor}(\text{R-}\underline{\text{free}})$)

respecting the objects (i.e. objects are mapped onto objects) will be called an f (resp. f_o) - quasi-section provided

$$\sigma' u = f \quad (\text{resp. } \sigma_o' u_o = f \ ).$$

Let σ' (resp. σ_o') be any f (resp. f_o) - quasi-section and consider the cochain $O(\sigma')$ (resp. $O(\sigma_o')$) of

$C^2((\text{S-}\underline{\text{free}}/_A)^o, \text{Der}_S(-, A \otimes I))$ (resp.

$C^2((\text{S-}\underline{\text{free}}/_A)^{epi,o}, \text{Der}_S(-, A \otimes I))$) defined by

$$O_o(\sigma')\left[\begin{array}{ccccc} F_o & \overset{\alpha_1}{\rightarrow} & F_1 & \overset{\alpha_2}{\rightarrow} & F_o \\ & \delta_o \searrow & \downarrow \delta_1 & \swarrow \delta_2 & \\ & & A & & \end{array}\right] = (\sigma'(\alpha_1\alpha_2) - \sigma'(\alpha_1)\sigma'(\alpha_2))(\delta_2 \otimes 1_I) \quad (\text{resp.}$$

$$O_o(\sigma_o')\left[\begin{array}{ccccc} F_o & \overset{\alpha_1}{\rightarrow} & F_1 & \overset{\alpha_2}{\rightarrow} & F_o \\ & \delta_o \searrow & \downarrow \delta_1 & \swarrow \delta_2 & \\ & & A & & \end{array}\right] = (\sigma_o'(\alpha_1\alpha_2) - \sigma'(\alpha_1)\sigma_o'(\alpha_2))(\delta_2 \otimes 1_I))$$

One proves as in (1.2) that $O_o(\sigma')$ (resp. $O_o(\sigma'_o)$) is a cocycle, and that the corresponding cohomology class coincides with the cohomology class $o(\pi,A)$ constructed above. Now suppose there exists a lifting A' of A to R. Then we may, for every object $(F_o \overset{\delta_o}{\to} A)$ of S-<u>free</u>$/_A$, pick an object $(F'_o \overset{\delta'_o}{\to} A')$ of R-<u>free</u>$/_{A'}$ such that $\delta'_o \otimes S = \delta_o$.

Obviously $\sigma'(\delta_o) = F'_o$, and let us put

$$\sigma'_{A'}(\delta_o) = \delta'_o .$$

With these notations let $Q_o = Q_o(\sigma',A')$ be the 1.cochain of $C^{\cdot}((\text{S-}\underline{\text{free}}/_A)^o, \mathrm{Der}_S(-,A\otimes I))$ defined by

$$Q_o\begin{pmatrix} F_o \overset{\alpha_1}{\to} F_1 \\ \delta_o \searrow \ \swarrow \delta_1 \\ A \end{pmatrix} = \sigma'(\alpha_1)\sigma'_{A'}(\delta_1) - \sigma'_{A'}(\delta_o) .$$

We find

$$(dQ_o)\begin{pmatrix} F_o \overset{\alpha_1}{\to} F_1 \overset{\alpha_2}{\to} F_2 \\ \delta_o \searrow \ \downarrow\delta_1 \ \swarrow \delta_2 \\ A \end{pmatrix} = \alpha_1(\sigma'(\alpha_2)\sigma'_{A'}(\delta_2) - \sigma'_{A'}(\delta_1))$$

$$- (\sigma'(\alpha_1\alpha_2)\sigma'_{A'}(\delta_2) - \sigma'_{A'}(\delta_o)) + (\sigma'(\alpha_1)\sigma'_{A'}(\delta_1) - \sigma'_{A'}(\delta_o))$$

$$= \sigma'(\alpha_1)(\sigma'(\alpha_2)\sigma'_{A'}(\delta_2) - \sigma'_{A'}(\delta_1)) - (\sigma'(\alpha_1\alpha_2)\sigma'_{A'}(\delta_2) - \sigma'_{A'}(\delta_o))$$

$$+ (\sigma'(\alpha_1)\sigma'_{A'}(\delta_1) - \sigma'_{A'}(\delta_o)) = (\sigma'(\alpha_1)\sigma'(\alpha_2) - \sigma'(\alpha_1\alpha_2))\sigma'_{A'}(\delta_2)$$

$$= -O_o(\sigma')(\alpha_1,\alpha_2) .$$

Thus $O_o(\sigma') = -dQ_o(\sigma',A')$ and $o(\pi,A) = 0$, proving the "if" part of the theorem.

Suppose $o(\pi,A) = 0$, then there exists a 1.-cochain ζ of $C^{\cdot}((S\text{-}\underline{free}/_A)^{o}, \mathrm{Der}_S(-, A \otimes I))$ such that $O(\sigma',A) = d\zeta$.

Since $j : D^{\cdot} \to C^{\cdot}$ is surjective there exists a 1-cochain ξ of $D^{\cdot}$ such that $j(\xi) = i\zeta$. Let σ_1 be given by

$$\sigma_1(\alpha) = \sigma'(f_o(\alpha)) + \xi(\alpha) .$$

Then σ_1 is an f_o-quasisection. (There is a slight difficulty at this point. One might find $\xi(1_{F_o}) \neq 0$. Observe however that $\zeta(1_{F_o}) = 0$ for all $\begin{matrix} F_o \\ \downarrow\delta_o \\ A \end{matrix}$. Therefore we may pick a ξ s.t. $\xi(1_{F_o}) = 0$, thus σ_1 is an f_o-quasisection.)

One checks that the 2.cochain ω of $D^{\cdot}$ defined by

$$\omega\left(\begin{matrix} F_o & \xrightarrow{\alpha_1} & F_1 & \xrightarrow{\alpha_2} & F_2 \\ & \delta_o\searrow & \downarrow\delta_1 & \swarrow\delta_2 & \\ & & A & & \end{matrix}\right) = \sigma_1(\alpha_1\alpha_2) - \sigma_1(\alpha_1)\sigma_1(\alpha_2)$$

is mapped to zero by j , thus sits in K^2 .

Now

$$A' = \varinjlim_{(S\text{-}\underline{free}/_A)^{epi}} \sigma_1 = \mathrm{coker}\Big(\coprod_{\left[\begin{smallmatrix} F_o & \xrightarrow{\alpha_1} & F_1 \\ \delta_o\searrow & & \swarrow\delta_1 \\ & A & \end{smallmatrix}\right]} \sigma_1(\delta_o) \rightrightarrows \coprod_{\left[\begin{smallmatrix} F \\ \downarrow\delta_o \\ A \end{smallmatrix}\right]} \sigma_1(\delta_o)\Big)$$

exists as an R-module. We shall show that A' is a lifting of A , thus justifying our claim of "good idea" above.

Consider the resolving complex $C. = \underline{C}.((S\text{-}\underline{free}/_A)^{epi},-)$ of $\varinjlim_{(S\text{-}\underline{free}/_A)^{epi}}$, for details see (La 1) or the Appendix.

Since σ_1 is not a functor $C.(\sigma_1)$ will not necessarily be a complex, but nevertheless we may consider the commutative diagram

$$\begin{array}{ccccccc}
0 & & 0 & & 0 & & A \otimes_S I \\
\downarrow & & \downarrow & & \downarrow & & \wr\! \| \\
C_2(\sigma_1) \otimes_R I & \dashrightarrow & C_1(\sigma_1) \otimes_R I & \to & C_o(\sigma_1) \otimes_R I & \xrightarrow{\beta} & A' \otimes_R I \to 0 \\
\downarrow & & \downarrow & & \downarrow & & \downarrow \alpha \\
C_2(\sigma_1) & \overset{\delta}{\dashrightarrow} & C_1(\sigma_1) & \overset{\gamma}{\to} & C_o(\sigma_1) & \to & A' \quad\; 0 \\
\downarrow & & \downarrow & & \downarrow & & \downarrow \\
C_2(\sigma_1) \otimes_R S & \to & C_1(\sigma_1) \otimes_R S & \to & C_o(\sigma_1) \otimes S & \to & A \to 0 \\
\downarrow & & \downarrow & & \downarrow & & \downarrow \\
0 & & 0 & & 0 & & 0
\end{array}$$

in which all sequences of morphisms marked with solid arrows are exact.

In fact we have $C.(\sigma_1) \otimes_R I = C.(g_o) \otimes_S I$ and $C.(\sigma_1) \otimes_R S = C.(g_o)$ where, we recall, $g_o = f_o i$ (see (2.2.2)). The vertical sequences are exact since all $C_p(\sigma_1)$ are R-free, the lower horizontal sequence is exact due to Corollary (2.2.2), and finally, part of the middle horizontal sequence is exact by the definition of A'.

Remember that we do not know that $\delta \circ \gamma = 0$. In fact it may well be that $\delta \circ \gamma \neq 0$. However $\mathrm{im}(\delta \circ \gamma) \subseteq C_o(\sigma_1) \otimes_R I$ and fortunately we have arranged the situation such that

$$\beta(\mathrm{im}(\delta \circ \gamma)) = 0 .$$

This follows by observing that the image of $\delta \circ \gamma$ consists of sums of elements of the form

$$(\sigma_1(\alpha_1 \alpha_2) - \sigma_1(\alpha_1)\sigma_1(\alpha_2))(\xi) = \omega(\alpha_1, \alpha_2)(\xi)$$

where

$$\begin{array}{ccccc}
F_0 & \xrightarrow{\alpha_1} & F_1 & \xrightarrow{\alpha_2} & F_2 \\
& {\scriptstyle \delta_o} \searrow & \downarrow {\scriptstyle \delta_1} & \swarrow {\scriptstyle \delta_2} & \\
& & A & &
\end{array}$$

are morphisms of $(S\text{-}\underline{\text{free}}_{/A})^{\text{epi}}$ and

$$\xi \in \sigma_1(\delta_o) = F'_0$$

Since $\omega \in K^2$ we conclude

$$\beta(\omega(\alpha_1,\alpha_2)(\xi)) = 0 \,.$$

Using this we may easily see that α is injective. But α is injective if and only if

$$\operatorname{Tor}_1^R(A',S) = 0 \,.$$

We have to show that A' is an R-algebra. Consider a system of homomorphisms

$$F_1 \xrightarrow{d} F_0 \underset{A}{\times} F_o \underset{p_2'}{\overset{p_1'}{\rightrightarrows}} F_0 \xrightarrow{\rho} A, \qquad \Delta' : F_o \to F_o \underset{A}{\times} F_o$$

in which F_0 and F_1 are free S-algebras, ρ and d are surjective, p_1' and p_2' are the projections and Δ' is the diagonal. Let $\Delta : F_0 \to F_1$ be a homomorphism such that $\Delta \circ d = \Delta'$, and put $p_i = d \circ p_i'$.

Then A is the inductive limit of the system

$$F_1 \underset{p_2}{\overset{p_1}{\rightrightarrows}} F_o, \qquad \Delta : F_o \to F_1$$

Apply the f_o-quasisection σ_1 on the corresponding morphisms of $(S\text{-}\underline{\text{free}}_{/A})^{\text{epi}}$. Then we get a diagram of R-algebras

$$F_1' \underset{\sigma_1(p_2)}{\overset{\sigma_1(p_1)}{\rightrightarrows}} F_o', \qquad \sigma_1(\Delta) : F_o' \to F_1'$$

Since we have the commutative diagram

$$
\begin{array}{ccccccccc}
0 & & 0 & & & & & & \\
\downarrow & & \downarrow & & & & & & \\
F_1 \otimes I & \rightrightarrows & F_0 \otimes I & \overset{\rho\otimes 1_I}{\rightarrow} & A \otimes I & \rightarrow & 0 & & \\
\downarrow & & \downarrow & & \downarrow & & & \searrow^{\alpha} & \\
F_1' & \overset{\sigma_1(p_1)}{\underset{\sigma_1(p_2)}{\rightrightarrows}} & F_0' & \underset{\rho'}{\rightarrow} & \operatorname{coker}(\sigma_1(p_1),\sigma_1(p_2)) & \overset{\beta}{\dashrightarrow} & & & A' \\
\downarrow & & \downarrow & & \downarrow & & & \swarrow & \\
F_1 & \overset{p_1}{\underset{p_2}{\rightrightarrows}} & F_0 & \underset{\rho}{\rightarrow} & A & \rightarrow & 0 & & \\
\downarrow & & \downarrow & & \downarrow & & & & \\
0 & & 0 & & 0 & & & &
\end{array}
$$

in which α is injective and all sequences involving morphisms marked with solid arrows are exact we conclude that β is an isomorphism. We are therefore through if we may prove that the R-module $\ker \rho' = \operatorname{im}(\sigma_1(p_1) - \sigma_1(p_2))$ is an ideal of F_0'.

Suppose $x \in \operatorname{im}(\sigma_1(p_1) - \sigma_1(p_2))$ and $y \in F_0'$. We have to prove that $yx \in \operatorname{im}(\sigma_1(p_1) - \sigma_1(p_2))$. First, assume $x \in F_0 \otimes I$, then $\rho'(yx) = (\rho\otimes 1)(\bar{y}\cdot x) = \rho(\bar{y})\cdot(\rho\otimes 1)(x) = 0$ where $\bar{y}$ is the image of y in F_0. Thus $yx \in \ker \rho'$. Since σ_1 is an f_0-quasisection we have

$$
\sigma_1\left(\begin{array}{ccc} F_0 & \overset{1_{F_0}}{\rightarrow} & F_0 \\ & \searrow \; \swarrow & \\ & A & \end{array}\right) = F' \overset{1_{F_0'}}{\rightarrow} F_0' .
$$

Therefore

$$
\sigma_1(\Delta)\sigma_1(p_i) = 1_{F_0'} - \omega(\Delta,p_i) \qquad i = 1,2 .
$$

We have already seen that for all $y \in F_0'$

$$
\omega(\Delta,p_i)(y) \in F_0 \otimes I
$$

$$
\omega(\Delta,p_i)(y) \in \ker \rho'
$$

Now

$$y = \sigma_1(p_i)(\sigma_1(\Delta)(y)) + \omega(\Delta, p_i)(y) \qquad i = 1,2$$

amd since $x \in \mathrm{im}(\sigma_1(p_1) - \sigma_1(p_2))$ there is a $u \in F_1'$ such that $x = \sigma_1(p_1)(u) - \sigma_1(p_2)(u)$

therefore

$$\begin{aligned} yx &= (\sigma_1(p_1)(\sigma_1(\Delta)(y)) + \omega(\Delta,p_1)(y))(\sigma_1(p_1)(u)) \\ &\quad - (\sigma_1(p_2)(\sigma_1(\Delta)(y)) + \omega(\Delta,p_2)(y))(\sigma_1(p_2)(u)) \\ &= \sigma_1(p_1)(\sigma_1(\Delta)(y)\cdot u) + \sigma_1(p_1)(u)\cdot\omega(\Delta,p_1)(y) \\ &\quad - \sigma_1(p_2)(\sigma_1(\Delta)(y)\cdot u) + \sigma_1(p_2)(u)\cdot\omega(\Delta,p_2)(y) . \end{aligned}$$

But, since we already know that

$$\sigma_1(p_i)(u)\cdot\omega(\Delta,p_i)(y) \in \ker \rho'$$

this shows that $yx \in \ker \rho' = \mathrm{im}(\sigma_1(p_1) - \sigma_1(p_2))$. Therefore A' is an algebra and we have proved that it is a lifting of A to R.

Fixing the quasisection σ', let A' be any lifting of A, and consider the 1.cochain $Q_o(\sigma',A')$ constructed above. Remember that $O(\sigma') = -dQ_o$. The corresponding σ_1 in the construction above, which is unique up to elements of K^1, will be denoted $\sigma'(A')$. For any object $(F_o \overset{\delta_o}{\to} A)$ of $(S\text{-}\underline{\text{free}}/_A)$ let us put $\sigma'(A')(\delta_o) = \sigma'_{A'}(\delta_o)$. For any morphism

$$\begin{array}{ccc} F_o & \overset{\alpha_1}{\to} & F_1 \\ & {}_{\delta_o}\searrow \quad \swarrow_{\delta_1} & \\ & A & \end{array}$$

of $(S\text{-}\underline{\text{free}}/_A)^{epi}$ we have, by definition of $Q_o(\sigma',A')$ a commutative diagram

$$\begin{array}{ccc} F'_o & \xrightarrow{\sigma'(A')(\alpha_1)} & F'_1 \\ & {\scriptstyle \sigma'(A')(\delta_o)} \searrow \quad \swarrow {\scriptstyle \sigma'(A')(\delta_1)} & \\ & A' & \end{array}$$

which implies

$$\lim_{\substack{\to \\ (S\text{-}\underline{free}/_A)^{epi}}} \sigma'(A') = A' .$$

Given a lifting A' there is thus a unique, up to elements of K^1, f_o-quasisection $\sigma'(A')$ such that

$$\lim_{\substack{\to \\ (S\text{-}\underline{free}/_A)^{epi}}} \sigma'(A') = A' .$$

Let A' and A'' be two liftings of A, then the corresponding cochain

$$Q_o(A') - Q_o(A'') \in C^1((S\text{-}\underline{free}/_A)^o, Der_S(-,A \otimes I))$$

is a cocycle defining a cohomology class

$$\lambda(A',A'') \in H^1(S,A;A \otimes I)$$

One easily checks that this class does not depend upon the choices made.
On the other hand if λ is an element of $H^1(S,A;A \otimes I)$, let ζ_o be a 1.cocycle of $C^{\cdot}((S\text{-}\underline{free}/_A)^o, Der_S(-,A \otimes I))$ representing λ, then we consider the 1.cochain

$$\zeta = Q_o(\sigma',A') - \zeta_o .$$

Obviously $d\zeta = -O(\sigma')$ and so there correspond ξ_o, $\xi \in D^1$ such that $j(\xi_o) = i(\zeta_o)$, $j(\xi) = i(\zeta)$. The quasisection $\sigma_1 = \sigma' + \xi$ defines a lifting A'' of A. One easily checks that

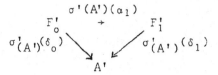

$$i(Q_o(A') - Q_o(A'')) = i(\zeta_o)$$

$$\sigma(A'') = \sigma(A') - \xi_o \qquad (\text{modulo } K^1)$$

Suppose $\lambda = 0$ then ζ_o is a coboundary. We may assume $\xi_o = d\eta$ with $\eta \in D^o$.

For every morphism

$$\begin{array}{ccc} F_o & \overset{\alpha_1}{\rightarrow} & F_1 \\ & {\scriptstyle \delta_o}\searrow \quad \swarrow{\scriptstyle \delta_1} & \\ & A & \end{array}$$

of $(S\text{-}\underline{free}/_A)^{epi}$ consider the diagram

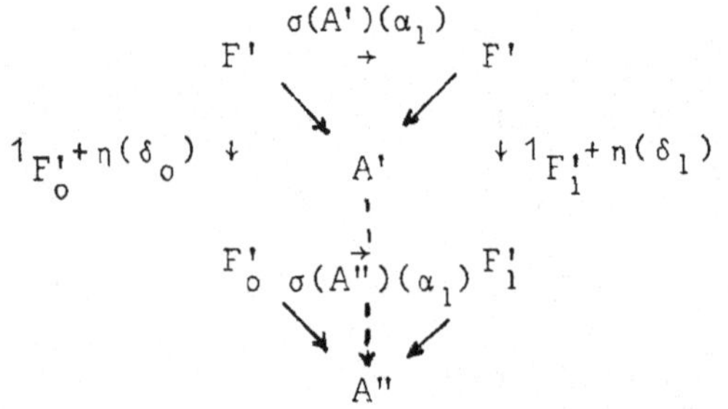

Since $i(\zeta_o) = d\eta$ we find that the diagrams of morphisms represented by solid arrows commute. But this implies the existence of a morphism $A' \rightarrow A''$ which joined to the solid diagrams will not distroy the commutativity. Obviously then $A'' \sim A'$. This proves the theorem. QED.

(2.3) Obstructions for lifting morphisms of algebras.

Let $\pi_1 : R_1 \rightarrow S_1$ and $\pi_2 : R_2 \rightarrow S_2$ be two surjective homomorphisms of commutative rings.

Let A_1 be an S_1-algebra and A_2 be an S_2-algebra and suppose given morphisms of rings β_o, β_1 and β_2 making the following diagram commutative:

$$\begin{array}{ccc} R_1 & \overset{\beta_0}{\to} & R_2 \\ \pi_1\downarrow & & \downarrow\pi_2 \\ S_1 & \overset{\beta_1}{\to} & S_2 \\ \mu_1\downarrow & & \downarrow\mu_2 \\ A_1 & \overset{\beta_2}{\to} & A_2 \end{array}$$

Suppose given a lifting A_1' of A_1 to R_1 and a lifting A_2' of A_2 to R_2.

<u>Definition (2.3.1)</u> A homomorphism of rings $\beta_2' : A_1' \to A_2'$ is a lifting of β_2 to β_o with respect to A_1' and A_2' if the following diagram commutes

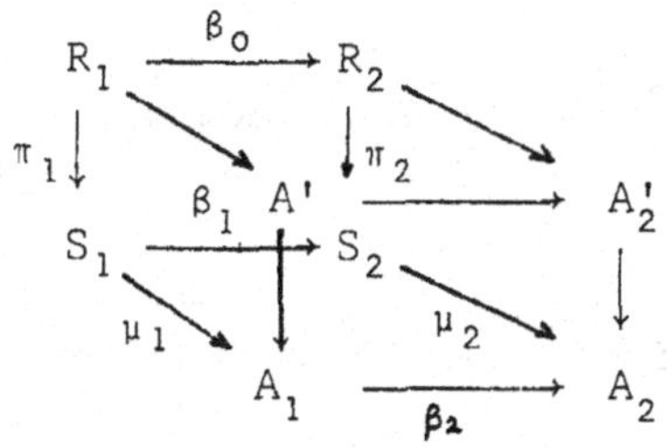

<u>Definition (2.3.2)</u> Two liftings β_2' and β_2'' of β_2 to β_o with respect to A_1' and A_2' are equivalent (written $\beta_2' \sim \beta_2''$) if there exist automorphisms of R-algebras $\theta_1 : A_1' \to A_1'$ and $\theta_2 : A_2' \to A_2'$ such that the following diagram commutes

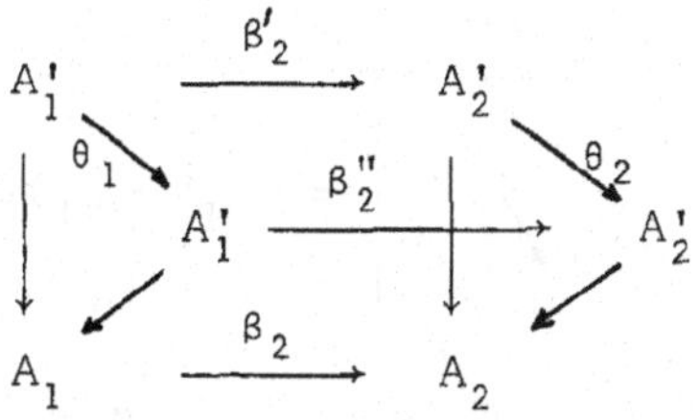

The set of liftings of β_2 to β_o w.r.t. A_1' and A_2' modulo this equivalence relation is called

$$\mathrm{Def}((\beta_o,\beta_1,\beta_2),A_1',A_2')$$

Now suppose $\ker \pi_1^2 = \ker \pi_2^2 = 0$. Then we may prove the following:

<u>Theorem (2.3.3)</u> Given liftings A_1' and A_2' of A_1 and A_2 respectively to R_1 and R_2 there exists an obstruction

$$o(\beta_2) = o(\beta_2,A_1',A_2') \in H^1(S_1;A_1;A_2 \underset{S_2}{\otimes} \ker \pi_2)$$

such that $o(\beta_2) = 0$ if and only if there exists a lifting of β_2 to β_o with respect to A_1' and A_2'. In this case $\mathrm{Def}((\beta_o,\beta_1,\beta_2),A_1',A_2')$ is a principal homogeneous space over

$$H^o(S_1,A_1;A_2 \otimes \ker \pi_2)/H^o(S_1,A_1,A_1 \otimes \ker \pi_1) + H^o(S_2,A_2,A_2 \otimes \ker \pi_2)$$

<u>Proof</u>. As in the proof of (2.2.5) pick any quasisection σ_i' of $- \underset{R_i}{\otimes} S_i : R_i$-<u>free</u> $\to S_i$-<u>free</u>, $i = 1,2$.

Consider the corresponding 1.cochain $Q_o(\sigma_i',A_i')$ of $C^{\cdot}((S_i\text{-}\underline{\text{free}}/_{A_i})^o, \mathrm{Der}_{S_i}(-,A_i \underset{S_i}{\otimes} \ker \pi_i))$ $i = 1,2$.

Let $O(\sigma_1',\sigma_2';A_1',A_2')$ be the 1.cochain of

$$C^{\cdot}((S_1\text{-}\underline{\text{free}}/_{A_1})^o, \mathrm{Der}_{S_1}(-,A_2 \underset{S_2}{o} \ker \pi_2))$$

defined by

$$O(\sigma_1',\sigma_2';A_1',A_2')\left(\begin{matrix} F_0 & \overset{\alpha_1}{\to} & F_1 \\ & {\scriptstyle\delta_o}\searrow \quad \swarrow{\scriptstyle\delta_1} & \\ & A_1 & \end{matrix}\right) = (1_{F_0} \otimes \beta_1)(\sigma_2'((\beta_o,\beta_1,\beta_2)*(\alpha_1))$$

$$- \sigma_1'(\alpha_1) \underset{R_1}{\otimes} 1_{R_2})(\beta_o,\beta_1,\beta_2)*(\delta_1) + Q_o(\sigma_1',A_1')(\alpha_1)(\beta_2 \underset{S_1}{\otimes} \beta_o)$$

$$- (1_{F_0} \underset{S_1}{\otimes} \beta_1)Q_o(\sigma_2',A_2')((\beta_o,\beta_1,\beta_2)*(\alpha_1))$$

where

$$(\beta_o,\beta_1,\beta_2)_* : (S_1\text{-}\underline{\text{free}}/_{A_1}) \to (S_2\text{-}\underline{\text{free}}/_{A_2})$$

is the functor defined by

$$(\beta_o,\beta_1,\beta_2)_* \begin{pmatrix} F_0 \\ \downarrow \delta_o \\ A_1 \end{pmatrix} = \begin{matrix} F_0 \underset{S}{\otimes} S_2 \\ \downarrow \delta'_o \\ A_2 \end{matrix} ,$$

δ'_o being the composition:

$$F_0 \underset{S_1}{\otimes} S_2 \to A_1 \underset{S_1}{\otimes} S_2 \to A_2 .$$

One checks that $O(\sigma'_1,\sigma'_2;A'_1,A'_2)$ is a 1.cocycle, and that the corresponding cohomology class $o(A'_1,A'_2) \in H^1(S_1,A_1 \underset{S_2}{\otimes} \ker \pi_2)$ does not depend upon the choices made.

Consider for every morphism of $(S\ \text{-}\underline{\text{free}}/_{A_1})$,

$$\begin{array}{ccc} F_o & \xrightarrow{\alpha_1} & F_1 \\ & {\scriptstyle\delta_o}\searrow \quad \swarrow{\scriptstyle\delta_1} & \\ & A_1 & \end{array}$$

the diagram:

$$\begin{array}{ccccccc} A'_1 & \xleftarrow{\sigma_{1A'_1}(\delta_1)} & F'_1 & \xrightarrow{1_{F'_1} \underset{R_1}{\otimes} \beta_o} & F'_1 \underset{R_1}{\otimes} R_2 & \xrightarrow{\sigma'_{2A'_2}(\delta'_1)} & A'_2 \\ {\scriptstyle\sigma_{1A'_1}(\delta_o)}\nwarrow & & \nearrow{\scriptstyle\sigma'_1(\alpha_1)} & & {\scriptstyle\sigma'_2(\alpha'_1)}\nwarrow & & \nearrow{\scriptstyle\sigma'_{2A'_2}(\delta'_o)} \\ & F'_o & & \xrightarrow[1_{F'_o} \underset{R_1}{\otimes} \beta_o]{} & & F'_o \underset{R_1}{\otimes} R_2 & \end{array}$$

where $\delta'_i = (\beta_o,\beta_1,\beta_2)_*(\delta_i)$, $i = 0,1$ and $\alpha'_1 = (\beta_o,\beta_1,\beta_2)_*(\alpha_1)$.

Put

$$\nu'(\delta_i) = (1_{F'_i} \otimes \beta_o)\sigma'_{2A'_2}(\delta'_i) \qquad i = 0,1 .$$

Then this diagram induces the following diagram

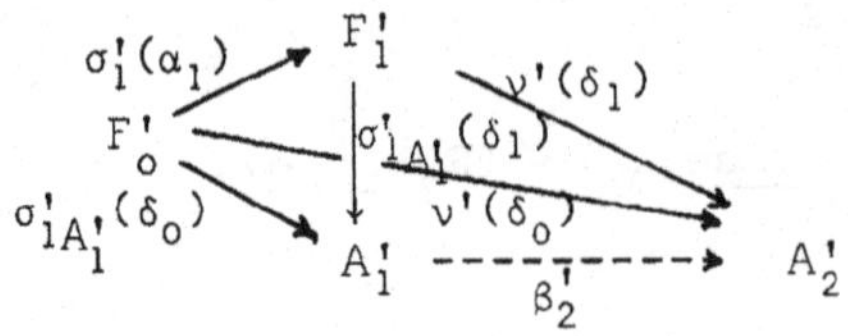

and we find that

$$O(\sigma_1',\sigma_2';A_1',A_2')\begin{bmatrix} F_0 & \overset{\alpha_1}{\rightarrow} & F_1 \\ \delta_o\searrow & & \swarrow\delta_1 \\ & A_1 & \end{bmatrix} = \nu'(\delta_o) - \sigma_1'(\alpha_1)\nu'(\delta_1)$$

$$+ Q_o(\sigma_1',A_1')(\alpha_1)(\beta_2 \underset{S_1}{\otimes} \beta_o)\,.$$

Suppose now that there exists a lifting β_2' of β_2 to β_o w.r.t. A_1' and A_2' then let $Q_1(\beta_2')$ be the 0.cochain of $C^\cdot((S_1\text{-}\underline{\text{free}}/_{A_1})^o, \mathrm{Der}_{S_1}(-,A_2 \underset{S_2}{\otimes} \ker \pi_2))$ defined by

$$Q_1(\beta_2')\begin{pmatrix} F_0 \\ \downarrow\delta_o \\ A \end{pmatrix} = \nu'(\delta_o) - \sigma_{1A'}'(\delta_o)\beta_2'\ .$$

We find

$$O(\sigma_1',\sigma_2';A_1',A_2') = -dQ_1$$

Thus proving the "if" part of the theorem.

Let us consider the image of $O(\sigma',A_1',A_2')$ in $C^\cdot$. By definition of $\sigma_1'(A_1')$ we find

$$\nu'(\delta_o) - \sigma_1'(\alpha_1)\nu'(\delta_1) + Q_o(\sigma_1',A_1')(\alpha_1)(\beta_2 \underset{S_1}{\otimes} \beta_o)$$

$$= \nu'(\delta_o) - \sigma_1'(A_1')(\alpha_1)\nu'(\delta_1)\ .$$

whenever

$$\begin{array}{ccc} F_0 & \overset{\alpha_1}{\rightarrow} & F_1 \\ \delta_o\searrow & & \swarrow\delta_1 \\ & A & \end{array}$$

in an object of $(A\text{-}\underline{\text{free}}/_A)^{epi}$.

With this done, suppose $o(A_1', A_2') = 0$, then there exists a 0.cochain κ of

$$C^{\cdot}((S_1\text{-}\underline{\text{free}}/_{A_1})^{o}, \mathrm{Der}_{S_1}(-, A_2 \underset{S_2}{\otimes} \ker \pi_2))$$

such that

$$d\kappa = O(\sigma_1', \sigma_2'; A_1', A_2') .$$

Put

$$\nu(\delta_o) = \nu'(\delta_o) - \kappa(\delta_o)$$

for every object

$$\begin{pmatrix} F_0 \\ \downarrow \delta_o \\ A_1 \end{pmatrix}$$

of $(S\text{-}\underline{\text{free}}/_{A_1})$, then for every morphism

$$\begin{array}{ccc} F_0 & \overset{\alpha_1}{\rightarrow} & F_1 \\ & \delta_o \searrow \quad \swarrow \delta_1 & \\ & A_1 & \end{array}$$

of $(S\text{-}\underline{\text{free}}/_{A})^{epi}$ we find a commutative diagram

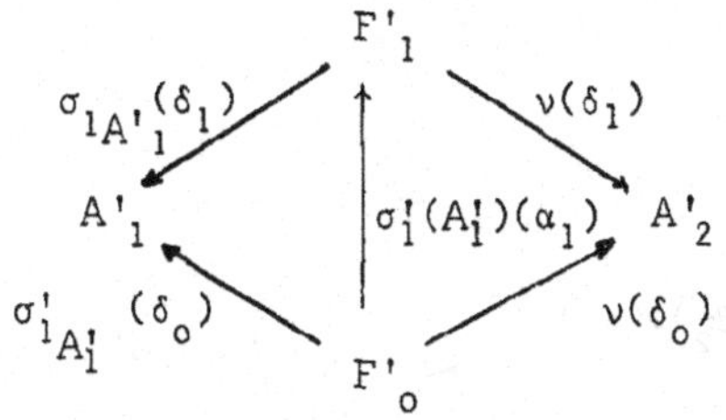

which proves that there exists a lifting β_2' of β_2 to β_o w.r.t. A_1' and A_2'. In fact we know that $A_1' = \varinjlim_{(S_1\text{-}\underline{\text{free}}/_{A_1})^{epi}} \sigma'(A_1')$.

This ends the proof of (2.3.3). QED.

<u>Remark (2.3.4)</u> By construction we have an equality

$\nu(\delta_o) = \sigma'_{1A'_i}(\delta_o)\beta'_2$ which implies that

$Q_1(\beta'_2)(\delta_o) = \nu'(\delta_o) - \sigma'_{1A'_1}(\delta_o)\beta'_2 = \kappa(\delta_o)$

for all objects of $(S_1\text{-}\underline{\text{free}}/_{A_1})$.

<u>Remark (2.3.5)</u> Consider any diagram of commutative rings

$$\begin{array}{ccccc} R_1 & \xrightarrow{\beta_o} & R_2 & \xrightarrow{\gamma_o} & R_3 \\ \pi_1\downarrow & & \pi_2\downarrow & & \pi_3\downarrow \\ S_1 & \xrightarrow{\beta_1} & S_2 & \xrightarrow{\gamma_1} & S_3 \\ \mu_1\downarrow & & \mu_2\downarrow & & \mu_3\downarrow \\ A & \xrightarrow{\beta_2} & A & \xrightarrow{\gamma_2} & A \end{array}$$

Assume π_1, π_2, π_3 are surjective and, moreover, that $\ker \pi_1^2 = \ker \pi_2^2 = \ker \pi_3^2 = 0$. Suppose given liftings A'_1, A'_2 and A'_3 of A_1, A_2 and A_3 respectively, and suppose we have found liftings β'_2, γ'_2, and $(\beta_2\gamma_2)'$ of β_2, γ_2 and $\beta_2\gamma_2$ respectively, w.r.t. A'_1 and A'_2, A'_2 and A'_3, and A'_1 and A'_3 respectively.

Pick any object $(F_0 \xrightarrow{\delta_o} A_1)$ of $S_1\text{-}\underline{\text{free}}/_{A_1}$ and consider the diagram

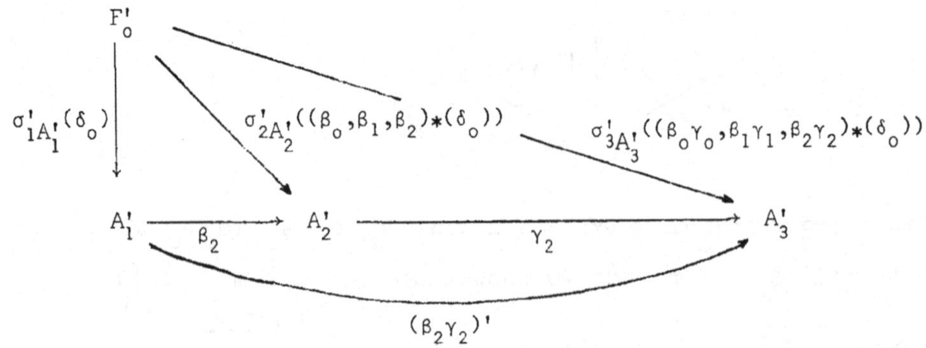

An easy computation shows that

$$\sigma'_1 A'_1(\delta_o)(\beta_2\gamma_2)' = \beta'_2\gamma'_2 - (Q_1(\gamma'_2)((\beta_o,\beta_1,\beta_2)*(\delta_o))$$

$$-Q_1((\beta_2\gamma_2)')(\delta_o) + Q_1(\beta'_2)(\delta_o)(\gamma_2 \otimes \gamma_o)) .$$

Chapter 3

Global cohomology

In this chapter we define a cohomology theory for any pair of categories $\underline{d}_o \subseteq \underline{d}$ of morphisms of algebras and any $\underline{d}$-Module F (see (3.1) for definitions). The idea is to fuse the André cohomology of algebras and the cohomology of categories defined in (1.1). The main point is the use of the category Mor $\underline{d}$.

We obtain a globalized André cohomology which we, for lack of better names, shall call the algebra cohomology of $(\underline{d}_o \subseteq \underline{d})$ with coefficients in F.

We then specialize to the case of schemes and morphisms of schemes, see (3.2). The main results of this section are the theorems (3.2.7), (3.2.9) and (3.2.11). We end the chapter by proving the existence of a long exact sequence of algebra cohomology associated to a morphism of schemes, generalizing the basic long exact sequence of the André cohomology.

(3.1) Definitions and some spectral sequences.

Let us first recall some fundamental constructions. If $\underline{e}$ is any small category, one may consider the category of functors on $\underline{e}$ with values in the category of abelian groups $\underline{Ab}$. We know that this category $\underline{Ab}^{\underline{e}}$ is abelian having enough injectives and projectives.

Let $C^{\cdot}(\underline{e},-) : \underline{Ab}^{\underline{e}} \to$ Compl.ab.gr. be the following functor:

Let F be any object of $\underline{Ab}^{\underline{e}}$ and put:

$$C^p(\underline{e},F) = \prod_{e_0 \underset{\mu_1}{\leftarrow} e_1 \underset{\mu_2}{\leftarrow} \cdots \underset{\mu_p}{\leftarrow} e_p} F(e_o)$$

where the indices run through all strings of p composable morphisms in $\underline{e}$. Let the differential $d^p : C^p(\underline{e},F) \to C^{p+1})\underline{e},F)$ be defined by: For $\beta = (\beta_{\mu_1,\mu_2,\ldots,\mu_p}) \in C^p(\underline{e},F)$ let $d^p(\beta) = (d^p(\beta)_{\mu_1,\mu_2,\ldots,\mu_{p+1}})$ be given by the formula

$$d^p(\beta)_{\mu_1,\mu_2,\ldots,\mu p+_1} = F(\mu_1)(\beta_{\mu_2,\mu_3,\ldots,{p+1}})$$

$$+ \sum_{i=1}^{p} (-1)^i \beta_{\mu_1,\ldots,\mu_i\mu_{i+1},\ldots,\mu_{p+1}} + (-1)^{p+1} \beta_{\mu_1,\mu_2,\ldots,\mu_p} .$$

It is easy to show that $d^p d^{p+1} = 0$ thereby proving that $C^\cdot(\underline{e},F) = \{C^p(\underline{e},F),d^p\}$ is a complex. Moreover, we observe that $C^\cdot(\underline{e},-)$ is an exact functor, and, almost by construction, we have (see Appendix or (La 1)) :

$$H^n(C^\cdot(\underline{e},-)) \simeq \underset{\overleftarrow{\underline{e}}}{\lim}{}^{(n)}$$

Given the category $\underline{e}$ we defined in (1.1) the category $\text{Mor}\ \underline{e}$ for which the objects are the morphisms of $\underline{e}$ and for which the morphisms $(\gamma,\delta) : \mu_1 \to \mu_2$ are commutative diagrams of the form

$$\begin{array}{ccc} e_1 & \underset{\gamma}{\leftarrow} & e_1' \\ \mu_1\downarrow & & \downarrow\mu_2 \\ e_2 & \underset{\delta}{\rightarrow} & e_2' \end{array}$$

For this special category we found another functorial complex $D^\cdot(\underline{e},-) : \underline{Ab}^{\text{Mor}\ \underline{e}} \to \underline{\text{Compl.ab.gr}}.$ with the same property as $C^\cdot(\text{Mor}\ \underline{e},-)$ but better suited for our purpose (see (1.3)). Recall that for an object G of $\underline{Ab}^{\text{Mor}\ \underline{e}}$ $D^\cdot(\underline{e},G)$ is given by:

$$D^p(\underline{e},G) = \prod_{e_o \xrightarrow{\mu_1} e_1 \xrightarrow{\mu_2} \cdots \xrightarrow{\mu_p} e_p} G(\mu_1 o \mu_2 \ldots o \mu_p)$$

and a differential

$$d^p : D^p(\underline{e},G) \to D^{p+1}(\underline{e},G) ,$$

where

$$d^p(\beta)_{\mu_1,\mu_2,\ldots,\mu_{p+1}} = G((\mu_1, 1_{e_{p+1}}))(\beta_{\mu_2,\ldots,\mu_{p+1}})$$

$$+ \sum_{i=1}^{p} (-1)^i \beta_{\mu_1,\ldots,\mu_i o \mu_{i+1},\ldots,\mu_{p+1}} + (-1)^{p+1} G((1_{e_o}, \mu_{p+1}))(\beta_{\mu_1,\ldots,\mu_p}) .$$

$D^{\cdot}(\underline{e},-)$ is an exact functor, and we proved in (1.3) that

$$H^n(D^{\cdot}(\underline{e},-) \simeq \varprojlim_{\text{Mor } \underline{e}}{}^{(n)} .$$

Recall also that for an object G of $\underline{Ab}^{\text{Mor } \underline{e}}$, we have defined the cohomology of $\underline{e}$ with coefficients in G by:

$$H^n(\underline{e},G) = H^n(D^{\cdot}(\underline{e},G)) = \varprojlim_{\text{Mor } \underline{e}}{}^{(n)} G .$$

We are now ready to start the construction of the algebra cohomology which eventually will lead to the cohomology groups $A^n(S,X;M)$ refered to in the Introduction.

Let S be any commutative ring with unit element, and let us make the following definition:

Definition (3.1.1) A 2.S-algebra is a morphism of S-algebras. If $\mu : A \to B$ and $\mu' : A' \to B'$ are 2.S-algebras, then a morphism $(\alpha,\beta) : \mu \to \mu'$ is a commutative diagram of the form

$$\begin{array}{ccc} A & \xrightarrow{\mu} & B \\ \alpha\downarrow & & \downarrow\beta \\ A' & \xrightarrow{\mu'} & B' \end{array}$$

Let 2.S-alg denote the category of 2.S-algebras, and consider a small subcategory $\underline{d}$ of 2.S-alg .

Obviously the functor $A \to (S \to A)$ defines an imbedding of S-alg in 2.S-alg . We shall therefore identify any small subcategory of S-alg with the corresponding subcategory of 2.S-alg .

Examples (3.1.2) (I) Let Y be any S-scheme and let $\mathfrak{U}$ be any affine open Zariski covering of Y . Then $\mathfrak{U}$ as a subset of the topology of Y is an ordered set, therefore a category, the morphisms being the inclusions $U \subseteq V$. The dual category $\underline{c}_{\mathfrak{U}}$ is a category of S-algebras.

(II) Let $f : X \to Y$ be a morphism of S-schemes, and let $\mathfrak{U}$ (resp. $\mathfrak{V}$) be an affine open covering of Y (resp. X). Then the set $f(\mathfrak{U}, \mathfrak{V}) = \{(U,V) \mid U \in \mathfrak{U}, V \in \mathfrak{V}, V \subseteq f^{-1}(U)\}$ is an ordered set and the dual category $\underline{d}_{\mathfrak{U},\mathfrak{V}}$ is a category of 2.S-algebras. In fact, if $(U,V) \in f(\mathfrak{U},\mathfrak{V})$ then $U = \mathrm{Spec}(A)$, $V = \mathrm{Spec}(B)$ and $f|V : V \to U$ corresponds to an S-algebra morphism $A \to B$.

Definition (3.1.3) A $\underline{d}$-Module M is a functor $M : \underline{d} \to \underline{Ab}$ such that for any object $\mu : A \to B$ of $\underline{d}$ $M(\mu)$ is a B-Module, and such that for every morphism $(\alpha,\beta) : \mu \to \mu'$ of $\underline{d}$ with $\mu' : A' \to B'$, the corresponding homomorphism $M((\alpha,\beta)) : M(\mu) \to M(\mu')$ is $\beta : B \to B'$ linear.

Example (3.1.4) In the situation of (3.1.2), (II) any O_X-Module will, in an obvious way, induce a $\underline{d}_{\mathfrak{U},\mathfrak{V}}$-Module.

Now, consider a morphism $(\alpha,\beta) : \mu \to \mu'$ of $\underline{d}$, i.e. a commutative diagram

$$\begin{array}{ccc} A & \xrightarrow[\mu]{} & B \\ \alpha\downarrow & & \downarrow\beta \\ A' & \xrightarrow[\mu']{} & B' \end{array}$$

Let M' be any B'-module and consider the functor

$$(\alpha,\beta)_* : A\text{-}\underline{\text{free}}/_B \to A'\text{-}\underline{\text{free}}/_{B'}$$

(see (2.2) for definitions) defined by:

$$(\alpha,\beta)_*(\delta) = \text{composition of } \delta \underset{A}{\otimes} 1_{A'} \text{ and } B \underset{A}{\otimes} A' \to B' .$$

where $\delta : A[\underline{x}] \to B$ denotes any object of $A\text{-}\underline{\text{free}}/_B$.
The functor $(\alpha,\beta)_*$ induces a morphism of complexes

$$C^{\cdot}((\alpha,\beta),M') : C^{\cdot}(A'\text{-}\underline{\text{free}}/_{B'}{}^{o},\text{Der}_{A'}(-,M')) \to C^{\cdot}(A\text{-}\underline{\text{free}}/_{B}{}^{o},\text{Der}_{A}(-,M')) .$$

If $\tau : M' \to M''$ is any homomorphism of B'-modules, then the corresponding homomorphisms of abelian groups

$$\tau_{\delta'} : \text{Der}_{A'}(A'[\underline{x}],M') \to \text{Der}_{A'}(A'[\underline{x}],M'')$$

where δ' runs through $A'\text{-}\underline{\text{free}}/_{B'}$ defines a morphism of functors

$$\tau : \text{Der}_{A'}(-,M') \to \text{Der}_{A'}(-,M'')$$

which in turn induces a morphism of complexes

$$C^{\cdot}(A'\text{-}\underline{\text{free}}/_{B'},\tau) : C^{\cdot}(A'\text{-}\underline{\text{free}}/_{B'}{}^{o},\text{Der}_{A'}(-,M')) \to C^{\cdot}(A'\text{-}\underline{\text{free}}/_{B'}{}^{o},\text{Der}_{A'}(-,M'')).$$

Let M be any d-Module then it follows from what has been said above that the map

$$(\alpha,\beta) \to C^{\cdot}(A\text{-}\underline{\text{free}}/_{B}{}^{o},\text{Der}_{A}(-,M(\mu')))$$

induces a functor

$$C^{\cdot}(-,\text{Der}_{-}(-,M)) : \text{Mor } \underline{d} \to \text{Compl.ab.gr.} .$$

We may therefore consider the double-complex

$$K_{\underline{d}}^{\cdot\cdot}(M) = D^{\cdot}(\underline{d}, C^{\cdot}(-,\mathrm{Der}_{-}(-,M))) .$$

Definition (3.1.5) The algebra cohomology of $\underline{d}$ with values in M, denoted by

$$A^n(S,\underline{d};M) = A^n(\underline{d};M) \quad n \geq 0 ,$$

is the cohomology of the simple complex associated to the double complex $K_{\underline{d}}^{\cdot\cdot}(M)$.

Remark (3.1.6) For $q \geq 0$ let $\underline{A}^q(M)$ denote the qth cohomology of the functor $C^{\cdot}(-,\mathrm{Der}_{-}(-,M))$, then $\underline{A}^q(M)$ is a functor on Mor $\underline{d}$ with values in $\underline{Ab}$.

Lemma (3.1.7) $A^n(S,\underline{d};M)$ is the abutment of a spectral sequence given by the term

$$E_2^{p,q} = H^p(\underline{d},\underline{A}^q(M)) .$$

Proof. This is just the first spectral sequence of $K_{\underline{d}}^{\cdot\cdot}$. QED

Let $\underline{d}_o$ be any subcategory of the category $\underline{d}$. Given a $\underline{d}$-Module M we may consider the restriction of M to Mor $\underline{d}_o$, which, abusing the language, will still be denoted by M. There is a canonical surjective morphism of double complexes

$$K_{\underline{d}}^{\cdot\cdot}(M) \to K_{\underline{d}_o}^{\cdot\cdot}(M)$$

Let $K_{\underline{d}/\underline{d}_o}^{\cdot\cdot}(M)$ denote the kernel of this morphism, and put:

Definition (3.1.8) The algebra cohomology of $\underline{d}$ relative to $\underline{d}_o$, with values in M, denoted by

$$A^n_{\underline{d}_o}(\underline{d},M) \qquad n \geq 0$$

is the cohomology of the simple complex associated to the double complex $K^{\cdot\cdot}_{\underline{d}/\underline{d}_o}(M)$.

There is a long exact sequence of cohomology

$$\cdots \to A^n_{\underline{d}_o}(\underline{d},M) \to A^n(\underline{d},M) \to A^n(\underline{d}_o,M) \to A^{n+1}_{\underline{d}_o}(\underline{d},M) \to \cdots$$

(3.2) Algebra cohomology of schemes and morphisms of schemes.

Now let us apply some of these generalities to algebraic geometry. First we have to suffer through some general nonsense expressed in the following lemmas.

Lemma (3.2.1) Let $\underline{e}$ be any small category, and consider the functor

$$\varepsilon : \text{Mor}\ \underline{e} \to \underline{e}$$

defined by:

$$\varepsilon(e_1 \to e_2) = e_2 .$$

Let $F : \underline{e} \to \underline{\text{Ab}}$ be any functor, then there are natural isomorphisms

$$\varprojlim_{\text{Mor}\ \underline{e}}{}^{(n)} \varepsilon F \simeq \varprojlim_{\underline{e}}{}^{(n)} F , \qquad n \geq 0$$

Proof. This is trivial, due to the fact that

$$D^{\cdot}(\underline{e},\varepsilon F) = C^{\cdot}(\underline{e},F) . \qquad \text{QED}$$

Corollary (3.2.2) Let $\underline{e}_o$ be any full subcategory of the category $\underline{e}$. Let $\underline{n}$ be the full subcategory of Mor $\underline{e}$ the objects of which are the morphisms $e_1 \to e_2$ with $e_2 \in \text{ob}.\,\underline{e}_0$. Suppose $\underline{e}$ has an initial object e_0 and let $F : \underline{e}_0 \to \underline{Ab}$ be any functor. Then there are natural isomorphisms

$$\varprojlim_{\underline{n}}{}^{(n)} \varepsilon F \simeq \varprojlim_{\underline{e}_0}{}^{(n)} F\,, \qquad n \geq 0\,.$$

Proof. Let $e_1 \to e_2$ be any object of $\underline{n}$, since e_0 is initial in $\underline{e}$ there is an isomorphism

$$\text{Mor}\begin{pmatrix} e_1 & e_0 \\ \downarrow\,, & \downarrow \\ e_2 & e \end{pmatrix} \simeq \text{Mor}(e_2, e)$$

for any object e of $\underline{e}$.

Let G be a Π-injective functor on $\underline{e}_0$ (see the Appendix for definition) defined by

$$G(e) = \prod_{\{e \to e_0' \,|\, e_0' \in \text{ob}\,\underline{e}_0\}} \bar{G}_{e_0'}$$

where $\{\bar{G}_{e_0'}\}_{e_0' \in \text{ob}\,\underline{e}_0}$ is a family of injective abelian groups. Then the restriction of $\varepsilon \circ G$ to $\underline{n}$ is also Π injective with

$$(\varepsilon \circ G)\begin{pmatrix} e_1 \\ \downarrow \\ e_2 \end{pmatrix} = G(e_2) = \prod_{\{e_2 \to e_0' \,|\, e_0' \in \text{ob}\,\underline{e}_0\}} \bar{G}_{e_0'} = \prod_{\left\{\begin{smallmatrix} e_1 & \to & e_1' \\ \downarrow & & \downarrow \\ e_2 & \to & e_2' \end{smallmatrix}\right\}} \bar{G}_{(e_1' \to e_2')}$$

where the family $\{\bar{G}_{(e_1' \to e_2')}\}_{(e_1' \to e_2') \in \text{ob}\,\underline{n}}$ is defined by

$\bar{G}_{(e_1' \to e_2')} = 0$ if $e_1' \neq e_0$, $\bar{G}_{(e_0 \to e_0')} = \bar{G}_{e_0'}$ if $e_0' \in \text{ob}\,\underline{e}_0$.

Since F has a resolution by Π-injective functors $G^{\cdot}$, the restriction of $\varepsilon \circ F$ to $\underline{n}$ has the Π-injective resolution $\varepsilon \circ G^{\cdot}$. Since $\varprojlim_{\underline{n}} \varepsilon \circ G \simeq \varprojlim_{\underline{e}_0} G$ for all functors G on $\underline{e}_0$ we find $\varprojlim_{\underline{n}}{}^{(\cdot)} \varepsilon \circ F \simeq \varprojlim_{\underline{e}_0}{}^{(\cdot)} F$. QED

Remark (3.2.3) The conclusion of the above corollary holds for any subcategory $\underline{n}$ of Mor $\underline{e}$ satisfying the following conditions

(1). $\varprojlim_{\underline{n}} \varepsilon \circ - \simeq \varprojlim_{\underline{e}_o} -$

(2) there is an object e_0 of $\underline{e}$ such that for any e'_0 of $\underline{e}_0$ there is an object $e_0 \to e'_0$ in $\underline{n}$ with

$$\mathrm{Mor}\left(\begin{matrix} e_1 \\ \downarrow \\ e_2 \end{matrix}, \begin{matrix} e_o \\ \downarrow \\ e_o \end{matrix}\right) \simeq \mathrm{Mor}(e_2, e'_0)$$

for all objects $e_1 \to e_2$ of $\underline{n}$.

Corollary (3.2.4) Let $\underline{e}$ be an ordered set and let $\underline{e}_0$ be any subset. Assume $\underline{e}$ has direct sums. Let $\underline{n}$ be the subset of the ordered set Mor $\underline{e}$ defined by

$$\mathrm{ob}\ \underline{n} = \{(e_1 \to e_2) \in \mathrm{ob\ Mor}\ \underline{e} \mid e_2 \in \underline{e}_0\}$$

Then there are isomorphisms

$$\varprojlim_{\underline{n}}{}^{(\cdot)} \varepsilon \circ F \simeq \varprojlim_{\underline{e}_0}{}^{(\cdot)} F$$

for all functors F on $\underline{e}_0$.

Proof. Consider the carrier function (κ-foncteur, see [La 3] p.245)

$$\kappa : \underline{e} \to \mathbb{P}\,\underline{n}$$

defined by

$$\kappa(e) = \{(e_1 \to e_2) \in \mathrm{ob}\ \underline{n} \mid e \to e_1\}$$

Notice that $\kappa(e)$ is cofinal in $\widehat{\kappa(e)}$ (see Appendix). In fact given an object y of $\widehat{\kappa(e)}$ such that there exist $x, x' \in \kappa(e)$ with $x \to y$, $x' \dot{\to} y$ there exists an $x_o \in \kappa(e)$

with

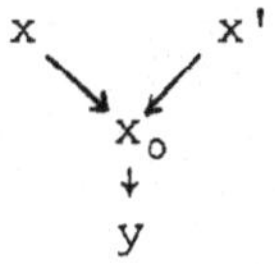

This follows easily from the diagram

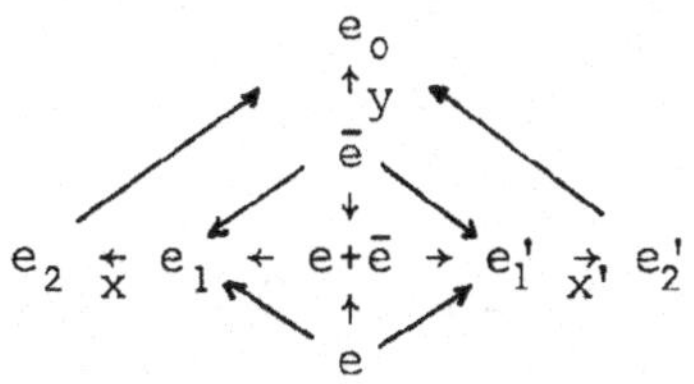

if we put $x_o = (e+\bar{e} \to e_o)$ (via the obvious compositions).

Notice also that $\kappa(e)$ is the category $\underline{n}$ of (3.2.2) corresponding to the subcategory $\hat{e} \cap \underline{e}_o$ of $\hat{e}$.

Let the carrier-function

$$\pi : \underline{e} \to P\underline{e}_o$$

be defined by

$$\pi(e) = \hat{e} \cap \underline{e}_o .$$

Then we have proved that the following canonical morphisms of complexes induce isomorphisms in cohomology:

$$C^{\cdot}(\underline{n},\varepsilon \circ F) \to C^{\cdot}(\underline{e};C^{\cdot}(\hat{\kappa};\varepsilon \circ F))$$
$$\to C^{\cdot}(\underline{e};C^{\cdot}(\kappa;\varepsilon \circ F)) \leftarrow C^{\cdot}(\underline{e},C^{\cdot}(\pi,F))$$
$$\leftarrow C^{\cdot}(\underline{e}_o,F),$$ (see [La 3] (1.3.1).) QED

Let X be an S-scheme, and let $\mathbb{Z}_X$ be the open covering of X consisting of all affine open subsets. Let $\underline{c}_X = \underline{c}_{\mathbb{Z}_X}$ (see (3.1.3) I.) be the dual category of S-algebras.

Let F be any O_X-Module then F is a $\underline{c}_X$-Module which, via the functor $\varepsilon : \mathrm{Mor}\,\underline{c}_X \to \underline{c}_X$ may be considered a functor on $\mathrm{Mor}\,\underline{c}_X$.

Theorem (3.2.5) Suppose F is quasicoherent, then there are natural isomorphisms

$$H^n(\underline{c}_X, F) \simeq H^n(X,F), \qquad n \geq 0 .$$

Proof. By (3.2.1) there are natural isomorphisms

$$H^n(\underline{c}_X, F) = \varprojlim_{\mathrm{Mor}\,\underline{c}_X}{}^{(n)} \varepsilon F \simeq \varprojlim_{\underline{\tilde{c}}_X}{}^{(n)} F .$$

Moreover we have isomorphisms

$$\varprojlim_{\underline{\tilde{c}}_X}{}^{(n)} F \simeq \varprojlim_{\mathbb{Z}^o_X}{}^{(n)} F \simeq H^n(X,F) . \qquad \text{QED}$$

Definition (3.2.6) The algebra cohomology of X with values in F are the groups

$$A^n(S,X;F) = A^n(S,\underline{c}_X;F), \qquad n \geq 0 .$$

If $\mu : A \to B$ is any object of $\mathrm{Mor}\,\underline{c}_X$, then (see (3.1.6))

$$\begin{aligned}\underline{A}^q(F)(\mu) &= H^q(C^{\cdot}(S\text{-}\underline{\mathrm{free}}/_A{}^o, \mathrm{Der}_S(-,F(\mathrm{Spec}(B))))) \\ &= H^q(S,A;F(\mathrm{Spec}(B))) .\end{aligned}$$

Using a result of André, (see (An) p. 85) we find that

$$H^q(S,A;F(\mathrm{Spec}(B))) = H^q(S,B;F(\mathrm{Spec}(B)))$$

and, in fact, $\underline{A}^q(F)$ is the composition of ε with a sheaf on X. This sheaf, which we shall still denote by $\underline{A}^q(F)$, is quasicoherent whenever F is.

Theorem (3.2.7) Suppose F is quasicoherent, then the algebra cohomology $A^{\cdot}(S,X;F)$ is the abutment of a spectral sequence given by the term

$$E_2^{p,q} = H^p(X,\underline{A}^q(F)) .$$

Proof. This is a trivial consequence of (3.2.5). QED

Consider any morphism of S-schemes $f : X \to Y$, and let $\mathbb{Z}_X$, and $\mathbb{Z}_Y$ be the ordered sets of affine open subsets of X and Y respectively. Put $\underline{d}_f = f(\mathbb{Z}_X, \mathbb{Z}_Y)^o$, see (3.1.2).

Let F be any O_X-Module. In an obvious way we may consider F a $\underline{d}_f$-Module.

Definition (3.2.8) The global algebra cohomology of f with values in F are the groups

$$A^n(f;F) = A^n(S,\underline{d}_f;F) , \qquad n \geq 0 .$$

Theorem (3.2.9) $A^\cdot(f;F)$ is the abutment of a spectral sequence with

$$E_2^{p,q} = H^p(\underline{c}_Y;\underline{A}_f^q(F))$$

where $A_f^q(F)$ is the functor on Mor $\underline{c}_Y$ defined by

$$\underline{A}_f^q(F)(A_1 \to A_2) = A^q(A_1, f^{-1}(\operatorname{Spec} A_2);F) .$$

Proof. Let $\tau : \text{Mor } \underline{d}_f \to \text{Mor } \underline{c}_Y$ be the functor defined by

$$\tau\begin{pmatrix} A & \longrightarrow & B \\ \downarrow & & \downarrow \\ A' & \longrightarrow & B' \end{pmatrix} = \begin{matrix} A \\ \downarrow \\ A' \end{matrix} .$$

Let $\pi : \text{Mor } \underline{c}_Y \to \mathbb{P}\, \text{Mor } \underline{d}_f$ be the carrier function defined by

$$\pi(x) = \tau^{-1}(\hat{x})$$

where $\hat{x} = \{x' \in \text{Mor } \underline{c}_Y \mid x \to x'\}$. Using [La 3](1.3.1) we find a homomorphism of complexes

$$C^\cdot(\text{Mor } \underline{d}_f,-) \to C^\cdot(\text{Mor } \underline{c}_Y, C^\cdot(\pi,-))$$

inducing isomorphisms in cohomology.

For x an object of Mor $\underline{c}_Y$ put $x = A \to A'$ and put

$\pi_o(x) =$

$$\left\{ \begin{matrix} A & \to & B \\ \downarrow & & \downarrow \\ A' & \to & B' \end{matrix} \in \pi(x) \,\middle|\, \exists\, A' \to B \text{ making the two resulting triangles commutative} \right\}$$

then $\pi_o(x)$ is cofinal in $\pi(x)$, (see the Appendix).

In fact, given any object

$$w = \begin{pmatrix} A_1 & \to & B_1 \\ \downarrow & & \downarrow \\ A_1' & \to & B_1' \end{pmatrix}$$

of $\pi(x)$, there exists by definition of $\pi(x)$ a commutative diagram

$$\begin{matrix} A & \leftarrow & A_1 \\ \downarrow & & \downarrow \\ A' & \to & A_1' \end{matrix} .$$

Put

$$u = \begin{pmatrix} A & \to & A' \otimes_{A_1} B_1 \\ \downarrow & & \downarrow \\ A' & \to & B_1' \end{pmatrix}$$

Then $u \in \pi_o(x)$ and the correspondence $w \mapsto u$ is a functor, projecting $\pi(x)$ onto $\pi_o(x)$.

This implies that $\Lambda_W(\pi_o(x))$ (see the Appendix) is acyclic, thus proving that the canonical homomorphism of complexes

$$C^\cdot(\pi,-) \to C^\cdot(\pi_o,-)$$

induces isomorphisms in cohomology.

Now, observe that $\pi_o(x)$ is isomorphic to Mor $\underline{c}_{f^{-1}(\mathrm{Spec}(A'))}$ where $f^{-1}(\mathrm{Spec}(A'))$ is considered as an A-scheme.

This implies that the cohomology of the double complex

$$C^\cdot(\pi\ (x), C^\cdot(-,\mathrm{Der}_-(-,F)))$$

being isomorphic to the cohomology of the double complex

$$C^{\cdot}(\pi_0(x),C^{\cdot}(-,\mathrm{Der}\_(-,F)) \simeq C^{\cdot}(\mathrm{Mor}\ \underline{c}_{f^{-1}(\mathrm{Spec}(A'))},C^{\cdot}(-,\mathrm{Der}\_(-,F)))$$

is equal to $\underline{A}^{\cdot}_{f}(F)(x)$.

Since we already know that the homomorphism of double complexes

$$C^{\cdot}(\mathrm{Mor}\ \underline{d}_f,C^{\cdot}(-,\mathrm{Der}\_(-,F))) \rightarrow$$

$$C^{\cdot}(\mathrm{Mor}\ \underline{c}_Y,C^{\cdot}(\pi,C^{\cdot}(-,\mathrm{Der}\_(-,F))))$$

induces isomorphisms in cohomology, the first spectral sequence of the double complex

$$C^{\cdot}(\mathrm{Mor}\ \underline{c}_Y,\overbrace{C^{\cdot}(\pi,C^{\cdot}}^{\cdot}(-,\mathrm{Der}\_(-,F))))$$

being given by

$$E_2^{p,q} = H^p(\underline{c}_Y;\underline{A}^q_f(F))$$

converges to the cohomology of $C^{\cdot}(\mathrm{Mor}\ \underline{d}_f,C^{\cdot}(-,\mathrm{Der}\_(-,F)))$ which is the same as the cohomology of the complex

$$D^{\cdot}(\underline{d}_f,C^{\cdot}(-,\mathrm{Der}\_(-,F))) .$$

This proves the theorem. QED

Consider a closed subscheme Z of the S-scheme X . The category $\underline{c}_{X-Z}$ is a ordered subset of the ordered set $\underline{c}_X$. Let F be any O_X-Module, and let's make the following definition.

<u>Definition (3.2.10)</u> The global algebra cohomology of X with values in F and support in Z , are the groups

$$A^n_Z(S,X;F) = A^n_{\underline{c}(X-Z)}(S,\underline{c}_X;F) \qquad n \geq 0 .$$

By construction we have a long exact sequence

$$\rightarrow A^n_Z(S,X;F) \rightarrow A^n(S,X;F) \rightarrow A^n(S,X-Z;F) \rightarrow A^{n+1}_Z(S,X;F) \rightarrow .$$

Let for any subset $\underline{e}_0$ of the ordered set $\underline{e}$,

$$C^{\bullet}(\underline{e}/\underline{e}_0,-)$$

denote the kernel of the canonical morphism

$$C^{\bullet}(\underline{e},-) \to C^{\bullet}(\underline{e}_0,-) .$$

Recall that we denote by $\hat{\underline{e}}_0$ the subset of $\underline{e}$ defined by

$$\hat{\underline{e}}_0 = \{x \in \underline{e} \mid \exists x' \in \underline{e}_0, x \leq x'\} .$$

By definition we have an exact sequence of double complexes

$$o \to C^{\bullet}(\mathrm{Mor}\ \underline{c}_X/\mathrm{Mor}\ \underline{c}_{X-Z}, C^{\bullet}(-,\mathrm{Der}\_(-,F))) \to$$

$$C^{\bullet}(\mathrm{Mor}\ \underline{c}_X, C^{\bullet}(-,\mathrm{Der}\_(-,F))) \to$$

$$C^{\bullet}(\mathrm{Mor}\ \underline{c}_{X-Z}, C^{\bullet}(-,\mathrm{Der}\_(-,F))) \to o ,$$

inducing the long exact sequence above.

Using Corollary (3.2.3) we may prove that the canonical morphism of double complexes

$$C^{\bullet}(\mathrm{Mor}\ \underline{c}_X/\widehat{\mathrm{Mor}\ \underline{c}_{X-Z}}, C^{\bullet}(-,\mathrm{Der}\_(-,F))) \to C^{\bullet}(\mathrm{Mor}\ \underline{c}_X/\mathrm{Mor}\ \underline{c}_{X-Z}, C^{\bullet}(-,\mathrm{Der}\_(-,F)))$$

induces isomorphisms in cohomology. In fact, consider the morphism of short exact sequences

$$\begin{array}{ccccccccc} o & \to & C^{\bullet}(\mathrm{Mor}\ \underline{c}_X/\widehat{\mathrm{Mor}\ \underline{c}_{X-Z}}; C^{\bullet}) & \to & C^{\bullet}(\mathrm{Mor}\ \underline{c}_X; C^{\bullet}) & \to & C^{\bullet}(\widehat{\mathrm{Mor}\ \underline{c}_{X-Z}}; C^{\bullet}) & \to & o \\ & & \downarrow & & \| & & \downarrow & & \\ o & \to & C^{\bullet}(\mathrm{Mor}\ \underline{c}_X/\mathrm{Mor}\ \underline{c}_{X-Z}; C^{\bullet}) & \to & C^{\bullet}(\mathrm{Mor}\ \underline{c}_X; C^{\bullet}) & \to & C^{\bullet}(\mathrm{Mor}\ \underline{c}_{X-Z}; C^{\bullet}) & \to & o \end{array}$$

Since by assumption X is a scheme, the intersection of two affine open subschemes is an affine open. This proves that $\underline{c}_X$ has finite direct sums. Then $\widehat{\mathrm{Mor}\ \underline{c}_{X-Z}}$ is the subcategory $\underline{n}$ corresponding to $\underline{c}_0 = \underline{c}_{X-Z}$, $\underline{c} = \underline{c}_X$ in (3.2.4).

Since $H^p(C^{\bullet}(-,\mathrm{Der}\_(-,F))) = \varepsilon \circ \underline{A}^p(F)$ as functors on $\underline{n}$ the conclusion of (3.2.4) implies that the right hand vertical

morphism induces an isomorphism of the first spectral sequences, thus is a quasiisomorphism.
The contention follows by the 5 - lemma.

Define the carrier functions

$$\pi_i : \operatorname{Mor} \underline{c}_X \to \mathbb{P} \operatorname{Mor} \underline{c}_X \qquad i = 1,2,$$

by

$$\pi_1(x) = \hat{x} \quad \text{and} \quad \pi_2(x) = \hat{x} \cap \widehat{\operatorname{Mor} \underline{c}_{X-Z}}.$$

By [La 3] (1.3) there is a canonical morphism of double complexes

$$C^{\cdot}(\operatorname{Mor} \underline{c}_X/\widehat{\operatorname{Mor} \underline{c}_{X-Z}}, C^{\cdot}(S\text{-}\underline{free}/-^{o}, \operatorname{Der}_S(-,F))) \to$$

$$C^{\cdot}(\operatorname{Mor} \underline{c}_X, C^{\cdot}(\pi_1/\pi_2, C^{\cdot}(S\text{-}\underline{free}/-^{o}, \operatorname{Der}_S(-,F))))$$

inducing isomorphisms in cohomology.
Let $x = A \to B$ be an object of $\operatorname{Mor} \underline{c}_X$ then for any object $x' = (A' \to B')$ of $\pi_1(x)$ there is a unique commutative diagram

$$\begin{array}{ccc} A & \to & B \\ \uparrow & & \downarrow \\ A' & \to & B' \end{array} .$$

Corresponding to this diagram there is a functor

$$S\text{-}\underline{free}/_{A'} \to S\text{-}\underline{free}/_{A}$$

which induces a morphism of complexes:
$C^{\cdot}(S\text{-}\underline{free}/_{A}{}^{o}, \operatorname{Der}_S(-,F(B)')) \to C^{\cdot}(S\text{-}\underline{free}/_{A'}{}^{o}, \operatorname{Der}_S(-,F(B)'))$.
In this way we obtain a morphism of functors on $\pi_1(x)$

$$C^{\cdot}(S\text{-}\underline{free}/_{A}{}^{o}, \operatorname{Der}_S(-,F(-))) \to C^{\cdot}(S\text{-}\underline{free}/-^{o}, \operatorname{Der}_S(-,F(-))).$$

We already know, see [An] p. 83, that this morphism induces isomorphisms in cohomology. Consequently the induced morphism of double complexes

$$C^{\cdot}(\pi_1(x)/\pi_2(x), C^{\cdot}(S\text{-}\underline{\text{free}}/_A{}^{\circ}, \mathrm{Der}_S(-,F))) \to$$

$$C^{\cdot}(\pi_1(x)/\pi_2(x), C^{\cdot}(S\text{-}\underline{\text{free}}/-{}^{\circ}, \mathrm{Der}_S(-,F)))$$

induces isomorphisms in cohomology.

Now there is a canonical isomorphism

$$C^{\cdot}(\pi_1(x)/\pi_2(x), C^{\cdot}(S\text{-}\underline{\text{free}}/_A{}^{\circ}, \mathrm{Der}_S(-,F))) \simeq$$

$$C^{\cdot}(S\text{-}\underline{\text{free}}/_A{}^{\circ}, \mathrm{Der}_S(-, C^{\cdot}(\pi_1(x)/\pi_2(x), F))) .$$

Putting things together we find a morphism of complexes

$$C^{\cdot}(\mathrm{Mor}\ \underline{c}_X/\mathrm{Mor}\ \underline{c}_{X-Z}, C^{\cdot}(S\text{-}\underline{\text{free}}/-{}^{\circ}, \mathrm{Der}_S(-,F))) \to$$

$$C^{\cdot}(\mathrm{Mor}\ \underline{c}_X, C^{\cdot}(S\text{-}\underline{\text{free}}/{}^{\circ}, \mathrm{Der}_S(-, C^{\cdot}(\pi_1/\pi_2, F))))$$

inducing isomorphisms in cohomology.

Consider the exact sequence of complexes

$$0 \to C^{\cdot}(\pi_1(x)/\pi_2(x), F) \to C^{\cdot}(\pi_1(x), F) \to C^{\cdot}(\pi_2(x), F) \to 0 .$$

Suppose F is quasicoherent, then obviously

$$H^q(C^{\cdot}(\pi_1(x), F)) = H^q(\mathrm{Spec}(B), F)$$

Moreover, since obviously the subset $\pi_2(x)_0 = \{(A' \to B') \in \pi_2(x) | A' = A\}$ is cofinal in $\pi_2(x)$, and since $\pi_2(x)_0$ satisfies the conditions of the Remark (3.2.3) we find:

$$H^q(C^{\cdot}(\pi_2(x), F)) = H^q(\mathrm{Spec}(B) - Z, F) .$$

Thus

$$H^q(C^{\cdot}(\pi_1(x)/\pi_2(x), F)) = \underline{H}^q_Z(F)(\mathrm{Spec}(B)) .$$

From this follows,

<u>Theorem (3.2.11)</u> $A_Z^{\cdot}(S,X;F)$ is the abutment of a spectral sequence given by

$$E_2^{p,q} = A^p(S,X;\underline{H}_Z^q(F)) .$$

<u>Proof.</u> Take the first spectral sequence of the double complex

$$\overbrace{C^{\cdot}(\mathrm{Mor}\ \underline{c}_X, C^{\cdot}}^{\bullet}(S\text{-}\underline{free}/-^{o}, \mathrm{Der}_S(-, C^{\cdot}(\pi_1/\pi_2, F)))) . \qquad \text{QED}$$

(3.3) <u>Long exact sequence associated to a morphism of S-schemes.</u>

Let $\psi : A \to B$ be a morphism of S-algebras. Then ψ induces a functor

$$\psi_* : S\text{-}\underline{alg}/_A \to S\text{-}\underline{alg}/_B .$$

Denote by $\underline{C}/\psi_*$ the category whose objects are the commutative diagrams of S-algebra morphisms

1)
$$\begin{array}{ccc} A_1 & \longrightarrow & B_1 \\ \delta_1 \downarrow & & \downarrow \delta_2 \\ A & \underset{\psi}{\longrightarrow} & B \end{array}$$

where δ_1 is an object of $S\text{-}\underline{alg}/_A$ and δ_2 is an object of $S\text{-}\underline{alg}/_B$.

A morphism of $\underline{C}/\psi_*$ is a pair of morphisms of $S\text{-}\underline{alg}/_A$ respectively $S\text{-}\underline{alg}/_B$ making all diagrams commutative.

Consider the forgetfull functors,

$$\Phi_1 : \underline{C}/_{\psi_*} \to S\text{-}\underline{alg}/_A$$

$$\Phi_2 : \underline{C}/_{\psi_*} \to S\text{-}\underline{alg}/_B$$

and the functor

$$\Phi_3 : \underline{C}/_{\psi_*} \to A\text{-}\underline{alg}/_B$$

defined by

$$\Phi_3 \begin{pmatrix} A_1 & \longrightarrow & B_1 \\ \delta_1 \downarrow & & \downarrow \delta_2 \\ A & \underset{\psi}{\longrightarrow} & B \end{pmatrix} = \begin{pmatrix} A \underset{A_1}{\otimes} B_1 \\ \downarrow \\ B \end{pmatrix} .$$

Let M be any B-module and define the functors

$$D_i(M) : \underline{C}/\psi_*^o \to \underline{Ab} \qquad i = 1,2,3 .$$

by
$$D_1(M) \begin{pmatrix} A_1 \to B_1 \\ \downarrow \quad \downarrow \\ A \to B \end{pmatrix} = \mathrm{Der}_S(A_1,M) = \mathrm{Der}_S(\Phi_1(-),M) \begin{pmatrix} A_1 \to B_1 \\ \downarrow \quad \downarrow \\ A \to B \end{pmatrix}$$

$$D_2(M) \begin{pmatrix} A_1 \to B_1 \\ \downarrow \quad \downarrow \\ A \to B \end{pmatrix} = \mathrm{Der}_S(B_1,M) = \mathrm{Der}_S(\Phi_2(-),M) \begin{pmatrix} A_1 \to B_1 \\ \downarrow \quad \downarrow \\ A \to B \end{pmatrix}$$

$$D_3(M) \begin{pmatrix} A_1 \to B_1 \\ \downarrow \quad \downarrow \\ A \to B \end{pmatrix} = \mathrm{Der}_{A_1}(B_1,M) = \mathrm{Der}_A(\Phi_3(-),M) \begin{pmatrix} A_1 \to B_1 \\ \downarrow \quad \downarrow \\ A \to B \end{pmatrix}$$

The resulting sequence of functors on $\underline{C}/_{\psi_*^o}$

2) $$o \to D_3(M) \to D_2(M) \to D_1(M) \to o$$

is left exact but not necessarily exact.

However, if we restrict to the full subcategory $S\text{-}\underline{free}/_\psi$ of $\underline{C}/_{\psi_*}$ defined by the objects of the form 1) where

(i) A_1 is S-free i.e. δ_1 is an object of $S\text{-}\underline{free}/_A$.

(ii) B_1 is A_1-free, thus in particular δ_2 is an object of $S\text{-}\underline{free}/_B$.

(iii) The morphism $A_1 \to B_1$ is the canonical morphism making B_1, a free A_1-algebra.

Then 2) becomes exact.

Observe that the functors Φ_i $i = 1,2,3$ restricted to $S\text{-}\underline{free}/_\psi$ yield functors:

$$\Phi_1 : S\text{-}\underline{\text{free}}/_{\psi} \to S\text{-}\underline{\text{free}}/_{A}$$

$$\Phi_2 : S\text{-}\underline{\text{free}}/_{\psi} \to S\text{-}\underline{\text{free}}/_{B}$$

$$\Phi_3 : S\text{-free}/_{\psi} \to A\text{-}\underline{\text{free}}/_{B} .$$

Therefore Φ_i , $i = 1,2,3$ induce morphisms of complexes

$$\Phi_1^* : C^\cdot((S\text{-}\underline{\text{free}}/_A)^o, \mathrm{Der}_S(-,M)) \to C^\cdot((S\text{-}\underline{\text{free}}/_{\psi})^o, D_1(M))$$

$$\Phi_2^* : C^\cdot((S\text{-}\underline{\text{free}}/_B)^o, \mathrm{Der}_S(-,M)) \to C^\cdot((S\text{-}\underline{\text{free}}/_{\psi})^o, D_2(M))$$

$$\Phi_3^* : C^\cdot((A\text{-}\underline{\text{free}}/_B)^o, \mathrm{Der}_A(-,M)) \to C^\cdot((S\text{-}\underline{\text{free}}/_{\psi})^o, D_3(M)) .$$

Moreover, the short exact sequence 2) induces a short exact sequence of complexes

3) $$o \to C^\cdot((S\text{-}\underline{\text{free}}/_{\psi})^o, D_3(M)) \to C^\cdot((S\text{-}\underline{\text{free}}/_{\psi})^o, D_2(M)) \to C^\cdot((S\text{-}\underline{\text{free}}/_{\psi})^o, D_1(M)) \to o .$$

<u>Lemma (3.3.1)</u> The morphisms Φ_i^* , $i = 1,2,3$ induce isomorphisms in cohomology.

<u>Proof</u>. Since $D_i(M) = \Phi_i \circ \mathrm{Der}_S(-,M)$ $i = 1,2,3$. we may apply the Corollary (2.1.7). In fact put $\underline{c} = \underline{C}/_{\psi_*}{}^o$, $\underline{d}$ = one of the categories $(S\text{-}\underline{\text{alg}}/_A)^o$, $(S\text{-}\underline{\text{alg}}/_B)^o$ or $(A\text{-}\underline{\text{alg}}/_B)^o$, $\underline{M}$ the subcategory $(S\text{-}\underline{\text{free}}/_{\psi})^o$ and $\underline{N}$ one of the subcategories $(S\text{-}\underline{\text{free}}/_A)^o$, $(S\text{-}\underline{\text{free}}/_B)^o$ or $(A\text{-}\underline{\text{free}}/_B)^o$ respectively.

We have to check the assumptions of (2.1.7). Obviously $\underline{c}$ and $\underline{d}$ have fibered products and it is easy to see that for all objects

$$c = \left\{\begin{matrix} A' & \to & B' \\ \downarrow & & \downarrow \\ A & \to & B \end{matrix}\right\}$$

of $\underline{C}/_{\psi_*}$ there exists an $\underline{M}$-epimorphism $m_o \to c$ with m_o an object of $\underline{M}$. It suffices to pick a commutative diagram of

S-algebras

$$\begin{array}{ccccc} F_1 & \longrightarrow & F_2 & & \\ & \searrow^{\delta_1} & & \searrow^{\delta_2} & \\ & & A' & \longrightarrow & B' \\ & & \downarrow & & \downarrow \\ & & A & \longrightarrow & B \end{array}$$

(with $F_1 \to A$ and $F_2 \to B$ the composite diagonal arrows)

such that

$$\begin{array}{ccc} F_1 & \longrightarrow & F_2 \\ \downarrow & & \downarrow \\ A & \longrightarrow & B \end{array}$$

is an object of $S\text{-}\underline{\text{free}}/_{\psi}$ and such that δ_1 and δ_2 are surjective.

By definition Φ_i transforms $\underline{M}$-epimorphisms into $\underline{N}$-epimorphisms. The only points to check are:

1) Φ_i preserves fibered powers of $\underline{M}$-epimorphisms.

2) Φ_i induces isomorphisms

$$\varprojlim_{(S\text{-}\underline{\text{free}}/_A)^o} \mathrm{Der}_S(-,M) \xrightarrow{\sim} \varprojlim_{(S\text{-}\underline{\text{free}}/_{\psi})^o} D_1(M) \qquad i = 1$$

$$\varprojlim_{(S\text{-}\underline{\text{free}}/_B)^o} \mathrm{Der}_S(-,M) \xrightarrow{\sim} \varprojlim_{(S\text{-}\underline{\text{free}}/_{\psi})^o} D_2(M) \qquad i = 2$$

$$\varprojlim_{(A\text{-}\underline{\text{free}}/_B)^o} \mathrm{Der}_A(-,M) \xrightarrow{\sim} \varprojlim_{(S\text{-}\underline{\text{free}}/_{\psi})^o} D_3(M) \qquad i = 3 .$$

1) is easily seen to hold for $i = 1.2$. 2) is more or less trivial. To see that 1) holds for $i = 3$ notice that

$m_p = m_o \times_c \cdots \times_c m_o$ is the diagram:

$$= \left(\begin{array}{ccc} F_1 \times_{A'} F_1 \times_{A'} \cdots \times_{A'} F_1 & \to & F_2 \times_{B'} F_2 \times_{B'} \cdots \times_{B'} F_2 \\ \downarrow & & \downarrow \\ A & \longrightarrow & B \end{array} \right)$$

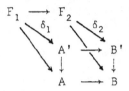

$n_o = \Phi_3(m_o)$ is the diagram $\begin{pmatrix} F_2 \underset{F_1}{\otimes} A \\ \downarrow \\ B \end{pmatrix}$

$$\Phi_3(m_p) = \begin{pmatrix} (F_2 \underset{B'}{\times} \cdots \underset{B'}{\times} F_2) \underset{(F_1 \underset{A'}{\times} \cdots \underset{A'}{\times} F_1)}{\otimes} A \\ \downarrow \\ B \end{pmatrix}$$

n_p is the diagram $\begin{pmatrix} (F_2 \underset{F_1}{\otimes} A) \underset{B'}{\times} \cdots \underset{B'}{\times} (F_2 \underset{F_1}{\otimes} A) \\ \downarrow \\ B \end{pmatrix}$

We may assume $F_1 \to A'$ (resp. $F_1 \to B'$) and $A' \to A$ (resp. $B' \to B$) both are surjective. Let $I = \ker(F_1 \to A')$.

We shall first prove

$$*) \quad (F_2 \underset{B'}{\times} \cdots \underset{B'}{\times} F_2) \underset{(F_1 \underset{A'}{\times} \cdots \underset{A'}{\times} F_1)}{\otimes} A' \simeq (F_2 \underset{F_1}{\times} A') \underset{B'}{\times} \cdots \underset{B'}{\times} (F_2 \underset{F_1}{\otimes} A').$$

This follows from

$$\ker(F_1 \underset{A'}{\times} \cdots \underset{A'}{\times} F_1 \to A') \simeq I \times \cdots \times I = I^{p+1}$$

Therefore the left side of *) is equal to

$$\operatorname{coker}(I^{p+1} \to F_2 \underset{B'}{\times} \cdots \underset{B'}{\times} F_2) \simeq \operatorname{coker}(I \to F_2) \underset{B'}{\times} \cdots \underset{B'}{\times} \operatorname{coker}(I \to F_2)$$

$$\simeq (F_2 \underset{F_1}{\otimes} A') \underset{B'}{\times} \cdots \underset{B'}{\times} (F_2 \underset{F_1}{\otimes} A').$$

Now, obviously

$$\begin{pmatrix} (F_2 \underset{B'}{\times} \cdots \underset{B'}{\times} F_2) \underset{(F_1 \underset{A'}{\times} \cdots \underset{A'}{\times} F_1)}{\otimes} A' \end{pmatrix} \underset{A'}{\otimes} A \simeq \Phi_3(m_p)$$

$$(F_2 \underset{F_1}{\otimes} A') \underset{A'}{\otimes} A \simeq F_2 \underset{F_1}{\otimes} A$$

We are therefore done if we can prove

$$**)\quad \{(F_2 \underset{F_1}{\otimes} A') \underset{B'}{\times} \cdots \underset{B'}{\times} (F_2 \underset{F_1}{\otimes} A')\} \underset{A'}{\otimes} A \simeq ((F_2 \otimes A') \underset{A'}{\otimes} A) \underset{B'\otimes_{A'} A}{\times} \cdots \underset{B'\otimes_{A'} A}{\times} ((F_2 \otimes A') \underset{A'}{\otimes} A)$$

This, however, follows from the following two lemmas

Lemma (3.3.2) Given a commutative ring A and an A-module M. Let $\underline{\Lambda}$ be any finite ordered set and let $F : \underline{\Lambda} \to A\text{-}\underline{\text{mod}}$ be any functor (projective system). Then there are two spectral sequences given by the terms:

$$'E_2^{p,q} = \mathrm{Tor}^A_{-p}(\underset{\underset{\underline{\Lambda}}{\rightarrow}}{\lim}{}^{(q)} F, M)$$

$$''E_2^{p,q} = \underset{\underset{\underline{\Lambda}}{\leftarrow}}{\lim}{}^{(p)} (\mathrm{Tor}^A_{-q}(F, M))$$

converging to the same graded A-module.

Proof. Since $\underline{\Lambda}$ is finite

$$C^{\cdot}(\underline{\Lambda}, F) \underset{A}{\otimes} M. \simeq C^{\cdot}(\underline{\Lambda}, F \underset{A}{\otimes} M.)$$

where $M.$ is a projective resolution of M as A-module. Moreover we may assume (by taking the subcomplex $C^{\cdot}_s$ of $C^{\cdot}$ given by the non-degenerated simplexes) that the corresponding double complex is bounded above, therefore the spectral sequences converge. QED.

Lemma (3.3.3) If $\underline{\Lambda}$ is of the form

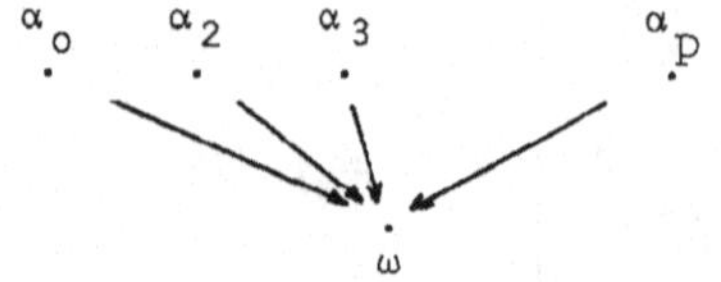

(i.e. corresponding to a fibered product)

and if $F : \Lambda \to A\text{-mod}$ is such that for all relations

$\alpha_i > \omega$, $F(\alpha_i > \omega)$ is surjective, then

$$\varprojlim_{\underline{\Lambda}}{}^{(i)} F = 0 \qquad \forall i \geq 1 .$$

<u>Proof</u>. This follows immediately from

$$\varprojlim_{\underline{\Lambda}}{}^{(i)} F = H^i(C_s^{\cdot}(\underline{\Lambda}, F))$$

(see [La 3]). QED.

Using these two lemmas, we observe that **) is a consequence of $F_2 \underset{F_1}{\otimes} A'$ being A'-<u>free</u>. QED.

Let $f: X \to Y$ be any morphism of S-schemes and consider any O_X-Module F. F induces a $\underline{d}_f$-Module, still denoted F. Consider the functor

$$C^{\cdot}(S\text{-}\underline{free}/-)^o, D_i(F(-))) : \mathrm{Mor}\ \underline{d}_f \to \underline{\mathrm{Compl.ab.gr.}}, \quad i = 1,2,3$$

defined by

$$(\psi \to \psi') = \begin{pmatrix} A & \to & A' \\ \downarrow\psi & & \psi'\downarrow \\ B & \to & B' \end{pmatrix} \to C^{\cdot}(S\text{-}\underline{free}/\psi)^o, D_i(F(\psi'))), \quad i = 1,2,3$$

Remember $F(\psi') = \Gamma(B') = F(\mathrm{Spec}(B'))$. The short exact sequence 3) above will induce a short exact sequence of functors $C^{\cdot}$ from which we deduce an exact sequence of double complexes:

$$\begin{aligned} ***) \qquad & 0 \to C^{\cdot}(\mathrm{Mor}\ \underline{d}_f, C^{\cdot}(S\text{-}\underline{free}/-, D_3(F(-)))) \\ & \to C^{\cdot}(\mathrm{Mor}\ \underline{d}_f, C^{\cdot}(S\text{-}\underline{free}/-, D_2(F(-)))) \\ & \to C^{\cdot}(\mathrm{Mor}\ \underline{d}_f, C^{\cdot}(S\text{-}\underline{free}/-, D_1(F(-)))) \to 0 \end{aligned}$$

With all this done, we shall state the main result of this paragraph:

Theorem (3.3.4) Let $f : X \to Y$ be any morphism of S-schemes, and consider any O_X-Module F. Then there is a long exact sequence

$$\cdots \to A^n(f;F) \to A^n(S,X;F) \to A^n(S,Y;R^{\cdot}f_*F) \to A^{n+1}(f;F) \to \cdots$$

where $A^n(S,Y;R^{\cdot}f_*F)$ denotes hypercohomology of the complex $R^{\cdot}f_*F$.

By (3.3.1) the following morphisms of double complexes induce isomorphisms in cohomology.

$$C^{\cdot}(\mathrm{Mor}\ \underline{d}_f, (p_1^o(-)\text{-}\underline{\mathrm{free}}/_{p_2^o(-)})^o, \mathrm{Der}_-(-, F(p_2^1(-))))$$
$$\downarrow r''$$
$$C^{\cdot}(\mathrm{Mor}\ \underline{d}_f, C^{\cdot}(S\text{-}\underline{\mathrm{free}}/_-)^o, D_3(F(-))))$$

$$C^{\cdot}(\mathrm{Mor}\ \underline{d}_f, C^{\cdot}(S\text{-}\underline{\mathrm{free}}/_{p_2^o(-)})^o, \mathrm{Der}_S(-, F(p_2^1(-))))$$
$$\downarrow$$
$$C^{\cdot}(\mathrm{Mor}\ \underline{d}_f, C^{\cdot}(S\text{-}\underline{\mathrm{free}}/_-)^o, D_2(F(-))))$$

$$C^{\cdot}(\mathrm{Mor}\ \underline{d}_f, C^{\cdot}(S\text{-}\underline{\mathrm{free}}/_{p_1^o(-)})^o, \mathrm{Der}_S(-, F(p_2^1(-))))$$
$$\downarrow$$
$$C^{\cdot}(\mathrm{Mor}\ \underline{d}_f, C^{\cdot}(S\text{-}\underline{\mathrm{free}}/_-)^o, D_1(F(-))))$$

where

$$p_i^j : \mathrm{Mor}\ \underline{d}_f \to S\text{-}\underline{\mathrm{alg}} \qquad i = 1,2\ ,\ j = 0,1$$

are the functors defined by:

$$p_1^o\begin{pmatrix} A & \to & B \\ \downarrow & & \downarrow \\ A' & \to & B' \end{pmatrix} = A \qquad p_1^1\begin{pmatrix} A & \to & B \\ \downarrow & & \downarrow \\ A' & \to & B' \end{pmatrix} = A'$$

$$p_2^o\begin{pmatrix} A & \to & B \\ \downarrow & & \downarrow \\ A' & \to & B' \end{pmatrix} = B \qquad p_2^1\begin{pmatrix} A & \to & B \\ \downarrow & & \downarrow \\ A' & \to & B' \end{pmatrix} = B'$$

Notice that p_i^o is contravariant and p_i^1 is covariant, $i = 1,2$.

Consider the functor

$$r : \operatorname{Mor} \underline{d}_f \to \operatorname{Mor} \underline{c}_X$$

defined by

$$r\begin{pmatrix} A & \to & B \\ \downarrow & & \downarrow \\ A' & \to & B' \end{pmatrix} = \begin{matrix} B \\ \downarrow \\ B' \end{matrix}$$

Then we have an equality of functors of complexes on $\operatorname{Mor} \underline{d}_f$

$$C^\cdot(S\text{-}\underline{\text{free}}/_{p_2^o}(-)^o, \operatorname{Der}_S(-, F(\psi_2^1(-)))$$

$$= r\, C^\cdot(S\text{-}\underline{\text{free}}/_-^o, \operatorname{Der}_S(-, F(-)))$$

This shows that there is a morphism of double complexes

$$\begin{array}{l} C^\cdot(\operatorname{Mor} \underline{c}_X, C^\cdot(S\text{-}\underline{\text{free}}/_-^o, \operatorname{Der}_S(-,F)))) \\ \downarrow 1'' \\ C^\cdot(\operatorname{Mor} \underline{d}_f, C^\cdot(S\text{-}\underline{\text{free}}/_{p_2^o}(-)^o, \operatorname{Der}_S(-, F(r(-)))))) \end{array}$$

We want to prove that $1''$ induces isomorphisms in cohomology. It suffices to prove that the corresponding morphism of the first spectral sequences involved is an isomorphism. I.e. we have to prove that the morphism

$$\begin{array}{lcl} H^p(C^\cdot(\operatorname{Mor} \underline{c}_X, \underline{A}^q(F))) & = & \varprojlim\limits_{\operatorname{Mor} \underline{c}_X}{}^{(p)} \underline{A}^q(F) \\ \downarrow & & \\ H^p(C^\cdot(\operatorname{Mor} \underline{d}_f, r\underline{A}^q(F))) & = & \varprojlim\limits_{\operatorname{Mor} \underline{d}_f}{}^{(p)} r\underline{A}^q(F) \end{array}$$

is an isomorphism for all p,q (see (3.1.7)). Now the functor $\underline{A}^q(F)$ on $\operatorname{Mor} \underline{c}_X$ is a sheaf on X, therefore (see (3.2.1))

$$\varprojlim\limits_{\operatorname{Mor} \underline{c}_X}{}^{(p)} \underline{A}^q(F) = \varprojlim\limits_{\underline{c}_X}{}^{(p)} \underline{A}^q(F)$$

$$\varprojlim\limits_{\operatorname{Mor} \underline{d}_f}{}^{(p)} \underline{A}^q(F) = \varprojlim\limits_{\underline{d}_f}{}^{(p)} \underline{A}^q(F)$$

To show that these groups are isomorphic under the given morphism is now nothing but a simple Leray spectral sequence argument for the morphism $f : X \to Y$. In fact, let G be any presheaf on X, let $\mathcal{V}$ (resp. $\mathcal{U}$) be any open covering of X (resp. Y) then consider the ordered set

$$f(\mathcal{U},\mathcal{V}) = \{(U,V) \mid U \in \mathcal{U}, V \in \mathcal{V}, V \subseteq f^{-1}(U)\}$$

and the order-preserving map

$$\psi : f(\mathcal{U},\mathcal{V}) \to \mathcal{V}$$

defined by $\psi(U,V) = V$. The image of ψ is an open covering of X, the intersection of $\mathcal{V}$ and $f^{-1}(\mathcal{U})$. Call it $\mathcal{T}$. Given $U \in \mathcal{U}$, let $\kappa(U)$ be the subset of $f(\mathcal{U},\mathcal{V})$ defined by:

$$\kappa(U) = \{(U',V') \mid U' \subseteq U, V' \subseteq f^{-1}(U')\}$$

It is easy to see that the subset

$$\kappa_0(U) = \{(U,V') \mid V' \subseteq f^{-1}(U)\}$$

is cofinal in $\kappa(U)$ (see the Appendix). Moreover we may identify $\kappa_0(U)$ with the subset $\{V' \in \mathcal{V} \mid V' \in f^{-1}(U)\} \subseteq \mathcal{V}$. With this identification we find:

$$\bigcup_{U \in \mathcal{U}^0} \kappa_0(U) = \mathcal{T}.$$

It then follows from ([La 3](1.3.1)) (see also the Appendix) that the two vertical and the lower horizontal morphisms of the following diagram

$$\begin{array}{ccc} C^{\cdot}(\mathcal{T},G) & \longrightarrow & C^{\cdot}(f(\mathcal{U},\mathcal{V}),G) \\ \downarrow & & \downarrow \\ C^{\cdot}(\mathcal{U},C^{\cdot}(\kappa_0(-),G)) & \longleftarrow & C^{\cdot}(\mathcal{U},C^{\cdot}(\kappa(-),G)) \end{array}$$

induce isomorphisms in cohomology. Put $\underline{c}_X = \mathcal{T}$, $\underline{d}_f = F(\mathcal{U},\mathcal{V})$ and $G = \underline{A}^q(F)$. Then the fact that the upper horizontal morphism of the diagram induces isomorphisms in cohomology proves that 1" induces isomorphisms in cohomology. To complete the proof of the theorem we shall compare the double complex

$$C^{\cdot}(\mathrm{Mor}\ \underline{d}_f, C^{\cdot}(S\text{-}\underline{\underline{\mathrm{free}}}/_{p_1^o}(-)^o, \mathrm{Der}_S(-, F(p_2^1(-)))))$$

and the following triple complex

$$C^{\cdot}(\mathrm{Mor}\ \underline{c}_Y, C^{\cdot}(S\text{-}\underline{\underline{\mathrm{free}}}/-^o, \mathrm{Der}_S(-, C^{\cdot}(\mathrm{Mor}\ \underline{c}_{f^{-1}(\mathrm{Spec}(-))}, F))))$$

In fact let us consider the obvious functor

$$h : \mathrm{Mor}\ \underline{d}_f \to \mathrm{Mor}\ \underline{c}_Y .$$

Given an object $A \to A'$ of $\mathrm{Mor}\ \underline{c}_Y$, put $\kappa(A \to A') = \widehat{h^{-1}(A \to A')}$. Then $\kappa(A \to A')$ is the full subcategory of $\mathrm{Mor}\ \underline{d}_f$ the objects of which are those diagrams

$$x = \begin{pmatrix} A_o & \to & B_o \\ \downarrow & & \downarrow \\ A'_o & \to & B'_o \end{pmatrix}$$

such that there exists a morphism

$$(A \to A') \to h(x) = (A_o \to A'_o)$$

of $\mathrm{Mor}\ \underline{c}_Y$, i.e. such that there is a commutative diagram

$$\begin{matrix} A_o & \to & A \\ \downarrow & & \downarrow \\ A'_o & \leftarrow & A' \end{matrix} .$$

Given such an x we find a commutative diagram

$$\begin{array}{ccccccc} B'_o & \leftarrow & B_o \otimes_{A_o} A' & \leftarrow & B_o \otimes_{A_o} A & \leftarrow & B_o \\ \uparrow & & \uparrow & & \uparrow & & \uparrow \\ A'_o & \leftarrow & A' & \leftarrow & A & \leftarrow & A_o \end{array}$$

From this follows that the full subcategory $\kappa_o(A \to A')$ of $\kappa(A \to A')$, the objects of which are the diagrams of the form:

$$\begin{array}{ccc} A & \to & B \\ \downarrow & \nearrow & \downarrow \\ A' & \to & B' \end{array}$$

is cofinal.

Since $C^{\cdot}(S\text{-}\underline{\text{free}}/_{p_1^o}(-)^o, \mathrm{Der}_S(-, F(p_2^1(-))))$ is a functor on $\mathrm{Mor}\ \underline{d}_f$, we find (see [La 3](1.3)) a morphism of double complexes

$$\begin{array}{l} C^{\cdot}(\mathrm{Mor}\ \underline{d}_f, C^{\cdot}(S\text{-}\underline{\text{free}}/_{p_1^o}(-)^o, \mathrm{Der}_S(-, F(p_2^1(-))))) \\ \downarrow \\ C^{\cdot}(\mathrm{Mor}\ \underline{c}_Y, C^{\cdot}(\kappa(-), C^{\cdot}(S\text{-}\underline{\text{free}}/_{p_1^o}(-)^o, \mathrm{Der}_S(-, F(p_2^1(-)))))) \end{array}$$

which induces isomorphisms in cohomology.

The morphism of double complexes

$$\begin{array}{l} C^{\cdot}(\kappa(A \to A'), C^{\cdot}(S\text{-}\underline{\text{free}}/_{p_1^o}(-)^o, \mathrm{Der}_S(-, F(p_2^1(-))))) \\ \downarrow \\ C^{\cdot}(\kappa_o(A \to A'), C^{\cdot}(S\text{-}\underline{\text{free}}/_{p_1^o}(-)^o, \mathrm{Der}_S(-, F(p_2^1(-))))) \\ \| \\ C^{\cdot}(S\text{-}\underline{\text{free}}/A^o, \mathrm{Der}_S(-, (, C^{\cdot}(\kappa_o(A \to A'), F(p_2^1(-)))))) \end{array}$$

induced by the inclusion $\kappa_o(A \to A') \subseteq \kappa(A \to A')$ induces isomorphisms in cohomology. Moreover it follows from the description of the objects of $\kappa_o(A \to A')$ that h maps $\kappa_o(A \to A')$

isomorphically (as ordered set, or category) onto

$$\mathrm{Mor}\ \underline{c}_{f^{-1}(\mathrm{Spec}(A'))}\ .$$

Thus, composing, we find a morphism of double complexes inducing isomorphisms in cohomology

$$\begin{array}{l} C^{\cdot}(\mathrm{Mor}\ \underline{d}_f, C^{\cdot}(S\text{-}\underline{\mathrm{free}}/_{p_1^o}(-)^o, \mathrm{Der}_S(-, F(p_2^1(-))))) \\ \downarrow \\ C^{\cdot}(\mathrm{Mor}\ \underline{c}_Y, C^{\cdot}(S\text{-}\underline{\mathrm{free}}/_{p^o}(-)^o, \mathrm{Der}_S(-, C^{\cdot}(\mathrm{Mor}\ \underline{c}_{f^{-1}(\mathrm{Spec}\, p^1(-))}, F)))) \end{array}$$

The conclusion of the theorem then follows from the exact sequence ***). QED.

Chapter 4

Global obstruction theory and formal moduli

In this chapter we first globalize the results of Chapter 2, adding what is necessary to treat the relativized version of the lifting problem.
Then we use the obstruction calculus to prove the main theorem of these notes, namely the structure theorem for the formal moduli of a pair of categories of morphisms of algebras, (4.2.4).

(4.1) Global obstruction theory

We shall define the notion of deformation of categories of 2.S-algebras in such a way that it generalizes the classical notion of infinitesimal deformations (liftings) of algebras and schemes, and moreover, takes care of the case of morphisms of schemes.
The applications we have in mind are many. We shall deduce results on moduli spaces, and in particular on the local structure of the Hilbert scheme. Hopefully we shall be able to use, in a later paper, the results of this chapter in the study of (possibly non-flat) descent.
This last application is partly responsible for the seemingly hopeless generalities that now follow.
Let 3.S-alg be the category in which the objects are the pairs of composable morphisms of S-alg, i.e. diagrams of the form $R \overset{\pi}{\to} A \overset{\mu}{\to} B$ in S-alg. If (π,μ) and (π',μ') are two objects of 3.S-alg then a morphism $(\pi,\mu) \to (\pi',\mu')$ of 3.S-alg is a tripple $(\beta_o,\beta_1,\beta_2)$ of morphisms of S-alg

making the following diagram commutative

$$\begin{array}{ccc} R & \xrightarrow{\beta_o} & R' \\ \pi\downarrow & & \downarrow\pi' \\ A & \xrightarrow[\beta_1]{} & A' \\ \mu\downarrow & & \downarrow\mu' \\ B & \xrightarrow[\beta_2]{} & B' \end{array}$$

Let

$$\Phi = \Phi_{1,3} : 3.S\text{-}\underline{alg} \to 2.S\text{-}\underline{alg}$$

be the functor defined by composition, i.e.

$$\Phi(R \to A \to B) = R \to B$$

and let

$$\Phi_{1,2} : 3.S\text{-}\underline{alg} \to 2.S\text{-}\underline{alg}$$

be the functor defined by $\Phi_{1,2}(R \to A \to B) = R \to A$.

<u>Definition (4.1.1)</u> Let $\underline{e}$ be any subcategory of $3.S\text{-}\underline{alg}$. A deformation of $\underline{e}$ is a functor σ making the following diagram commutative,

$$\begin{array}{ccc} \underline{e} & \xrightarrow{\sigma} & 3.S\text{-}\underline{alg} \\ & {}_{\Phi|\underline{e}}\searrow \quad \swarrow_{\Phi} & \\ & 2.S\text{-}\underline{alg} & \end{array}$$

such that for every object $(R \to A \to B)$ of $\underline{e}$ writing $\sigma(R \to A \to B) = R \to \sigma(B) \to B$ the following two conditions hold:

(1) $\sigma(B) \underset{R}{\otimes} A = B$

(2) $\mathrm{Tor}_1^R(\sigma(B), A) = 0$.

<u>Definition (4.1.2)</u> Two deformations σ and σ' of $\underline{e}$ are equivalent (written $\sigma \sim \sigma'$) if there is an isomorphism of

functors

$$\theta : \sigma \to \sigma'$$

such that $\Phi(\theta)$ is the identity on $\Phi|\underline{e}$.

Remark (4.1.3) It is easy to see that $\sim$ defines an equivalence relation in the set of deformations of $\underline{e}$ (N.B. we shall prefer not to enter into any set theoretical considerations at this point. See the Introduction.)

Definition (4.1.4) Let $\underline{e}$ be any small subcategory of 3.S-alg, then we shall denote by

$$\mathrm{Def}(\underline{e})$$

the set of deformations of $\underline{e}$ modulo the equivalence relation $\sim$ defined above.

Remark (4.1.5) Abusing the language we shall sometimes use the notation σ both for a deformation of $\underline{e}$ and for its equivalence class, hoping that this will simplify the exposition without introducing too much confusion.

Let $\underline{e}_0$ be any subcategory of $\underline{e}$. Then the inclusion $\underline{e}_0 \subseteq \underline{e}$ induces a canonical map

$$\mathrm{Def}(\underline{e}) \to \mathrm{Def}(\underline{e}_0) .$$

In fact, we may consider Def as a functor on the ordered set (category) of small subcategories of 3.S-alg .

Remark (4.1.6) Let $\underline{e}$ be such that for every object $R \overset{\pi}{\to} A \to B$ of $\underline{e}$ the morphism π is surjective, then a deformation of $\underline{e}$ will be refered to as a lifting of e .

Example (4.1.7) Let A be any S-algebra, and let $\pi : R \to S$ be any surjective morphism of commutative rings. Let $\underline{e}$ be the subcategory of $3.\mathbb{Z}$-alg consisting of the single object $(R \to S \to A)$ and the identity morphism. A deformation σ of $\underline{e}$ is then an R-algebra $\sigma(A)$ together with a morphism $\sigma(A) \to A$ such that the following conditions hold:

1. The diagram

$$\begin{array}{ccc} R & \longrightarrow & \sigma(A) \\ \downarrow & & \downarrow \\ S & \longrightarrow & A \end{array}$$

is commutative.

2. $\sigma(A) \underset{R}{\otimes} S \tilde{\to} A$.

3. $\mathrm{Tor}_1^R(\sigma(A),S) = 0$.

Thus a deformation (lifting) of $\underline{e}$ is simply a lifting of the A-algebra A to R .

Example (4.1.8) Let X be an S-scheme, and consider the category of S-algebras $\underline{c}_X$ (see (3.1)). Let $\pi : R \to S$ be any homomorphism of commutative rings and consider the subcategory $\underline{e}$ of $3.\mathbb{Z}$-alg the objects of which are the pairs of morphisms of $\mathbb{Z}$-algebras

$$R \to S \to A$$

where $S \to A$ is the structure morphism of an object of $\underline{c}_X$, the morphisms of $\underline{e}$ being the morphisms of $\underline{c}_X$ extended in the obvious way.

If π is surjective then a deformation (lifting) of $\underline{e}$

is a section σ of the functor

$$-\underset{R}{\otimes} S : R\text{-}\underline{alg} \to S\text{-}\underline{alg}$$

defined on the subcategory $\underline{c}_X$ of S-$\underline{alg}$, such that for any object A of $\underline{c}_X$ $\sigma(A)$ is a lifting of A to R. Suppose π has nilpotent kernel, i.e. that for some $n, (\ker \pi)^n = 0$, then σ corresponds to an R-scheme X' which is a deformation of X to R. In fact for each affine open subset Spec(A) of X we take Spec(σ(A)) and we glue. This sets up a one-to-one correspondence between the set of deformations of $\underline{e}$ and the set of deformations of the scheme X to R.

Example (4.1.9) Let $f : X \to Y$ be any morphism of S-schemes, and consider the category $\underline{d}_f$ of 2.S-algebras (see (3.1)). Let $\pi : R \to S$ be any morphism of S-algebras, and consider the subcategory $\underline{e}$ of 3.S-$\underline{alg}$ the objects of which are the pairs of morphisms

$$A \underset{S}{\otimes} R \xrightarrow[1_A \otimes \pi]{} A \underset{S}{\otimes} S = A \xrightarrow[\mu]{} B$$

μ running through the set of objects of $\underline{d}_f$, the morphisms being the morphisms of $\underline{d}_f$ extended in the obvous way. Suppose π is surjective and has nilpotent kernel, then a deformation σ of $\underline{e}$ corresponds to pairs of morphisms of S-schemes ε and f' making the following diagram cartesian

$$\begin{array}{ccc} X' & \xrightarrow{f'} & Y \underset{S}{\times} \mathrm{Spec}(R) \\ \varepsilon\uparrow & & \uparrow 1 \times \mathrm{Spec}(\pi) \\ X & \xrightarrow{f} & Y \end{array}$$

and satisfying the following condition

$$\mathrm{Tor}_1^{O_Y \otimes_S R}(O_{Y'}, O_Y) = 0$$

which reduces to

$$\mathrm{Tor}_1^R(O_{X'}, S) = 0 .$$

In fact, let Spec(A) be any open affine subset of Y, let Spec(B) be any open affine subset of X contained in $f^{-1}(\mathrm{Spec}(A))$, then $\sigma(A \otimes_S R \to A \overset{\mu}{\to} B) = A \otimes_S R \overset{\mu'}{\to} \sigma(B) \overset{\xi}{\to} B$. We may glue the $\mathrm{Spec}(\sigma(B))$'s together to form a scheme X'. The morphisms ε and f' correspond to the morphisms ξ and μ' respectively.
This sets up a one-to-one correspondence between the set of deformations of $\underline{e}$ and the set of deformations of f to R.

Now, suppose $\underline{e}$ is any small subcategory of 3.S-alg such that for any object $R \overset{\pi}{\to} A \overset{\mu}{\to} B$ of $\underline{e}$ the morphism π is surjective and $(\ker \pi)^2 = 0$.

Then $\ker \pi$ is an A-module, and the correspondence

$$(R \to A \to B) \to B \underset{A}{\otimes} \ker \pi$$

defines a functor

$$O \otimes \ker \Phi_{1,2} : \underline{e} \to \underline{\mathrm{Ab}}$$

which is an $\underline{e}$-Module.
We shall construct a functor

$$C^{\cdot}(-, \mathrm{Der}\_(-, O \otimes \ker \Phi_{1,2})) : \mathrm{Mor}\ \underline{e} \to \underline{\mathrm{Compl.ab.gr}}.$$

analogous to the functor

$$C^{\cdot}(-, \mathrm{Der}\_(-, M)) : \mathrm{Mor}\ \underline{d} \to \underline{\mathrm{Compl.ab.gr}}.$$

studied in (3.1).

Let $(\beta_o,\beta_1,\beta_2)$ be an object of Mor $\underline{e}$, i.e. a morphism $(\pi,\mu) \to (\pi',\mu')$ of $\underline{e}$, corresponding to the commutative diagram

$$\begin{array}{ccc} R & \xrightarrow{\quad} & R' \\ \pi\big\downarrow & \beta_o & \big\downarrow\pi' \\ A & \xrightarrow{\quad} & A' \\ \mu\big\downarrow & \beta_1 & \big\downarrow\mu' \\ B & \xrightarrow[\beta_2]{\quad} & B' \end{array}$$

Exactly as before we may convince ourselves that the correspondence

$$(\beta_o,\beta_1,\beta_2) \to C^{\cdot}((A\text{-}\underline{\text{free}}/_B)^o, \mathrm{Der}_A(-,B' \underset{A'}{\otimes} \ker \pi'))$$

defines a functor Mor $\underline{e} \to \underline{\text{Compl.ab.gr.}}$. This is the functor $C^{\cdot}(-,\mathrm{Der}_{-}(-,O \otimes \ker \Phi_{1,2}))$.

Consider the double complex

$$K_{\underline{e}}^{\cdot\cdot} = D^{\cdot}(\underline{e}, C^{\cdot}(-,\mathrm{Der}_{-}(-,O \otimes \ker \Phi_{1,2}))) .$$

<u>Definition (4.1.10)</u> We shall denote by

$$A^n(\underline{e},O) \qquad n \geq 0 ,$$

the cohomology of the simple complex associated to the double complex $K_{\underline{e}}^{\cdot\cdot}$.

<u>Examples (4.1.11)</u> In the situation of (4.1.7) there are canonical isomorphisms

$$A^n(\underline{e},O) \simeq H^n(S,A;A \underset{S}{\otimes} \ker \pi) \qquad n \geq 0 ,$$

provided $(\ker \pi)^2 = 0$.

In the situation of (4.1.8) there are canonical isomorphi[sms]

$$A^n(\underline{e},O) \simeq A^n(S,X;O_X \underset{S}{\otimes} \ker \pi) \qquad n \geq 0 ,$$

provided $(\ker \pi)^2 = 0$.

In the situation of (4.1.9) there are canonical isomorphisms

$$A^n(\underline{e},0) \simeq A^n(f;0_X \underset{S}{\otimes} \ker\pi) \qquad n \geq 0 .$$

In fact, the category $\underline{e}$ in these three cases is isomorphic to $\{S \to A\}$, $\underline{c}_X$ and $\underline{d}_f$ respectively.

Let $\underline{e}_0$ be any subcategory of the category $\underline{e}$. Then there is a canonical epimorphism of double complexes

$$K^{\cdot\cdot}_{\underline{e}} \to K^{\cdot\cdot}_{\underline{e}_0}$$

Let $K^{\cdot\cdot}_{\underline{e}/\underline{e}_0}$ be the kernel of this morphism.

Definition (4.1.12) We shall denote by

$$A^n_{\underline{e}_o}(\underline{e},0) \qquad n \geq 0$$

the cohomology of the simple complex associated to the double complex $K^{\cdot\cdot}_{\underline{e}/\underline{e}_0}$.

Thus, by definition, there is a long exact sequence of cohomology

$$\cdots \to A^n_{\underline{e}_o}(\underline{e},0) \to A^n(\underline{e},0) \to A^n(\underline{e}_0,0) \to A^{n+1}_{\underline{e}_o}(\underline{e},0) \to \cdots$$

Example (4.1.13) In the situation of (4.1.8) let X_o be a closed subscheme of X. Consider the subcategory $\underline{c}_{X-X_o}$ of $\underline{c}_X$ and the corresponding subcategory $\underline{e}_0$ of $\underline{e}$, then there are canonical isomorphisms

$$A^n_{\underline{e}_o}(\underline{e},0) \simeq A^n_{X_o}(S,X,0_X \underset{S}{\otimes} \ker\pi) \qquad n \geq 0 .$$

Theorem (4.1.14) There is an obstruction

$$o(\underline{e}) \in A^2(\underline{e},0)$$

such that $o(\underline{e}) = 0$ is a necessary and sufficient condition for the existence of a deformation of $\underline{e}$. If $o(\underline{e}) = 0$ then $\mathrm{Def}(\underline{e})$ is a principal homogenous space over $A^1(\underline{e},0)$

Proof. We shall start by constructing a 2-cocycle of the simple complex associated to the double complex $K_o^{\cdot\cdot}$, defining the cohomology class o. Using results of Chapter 2, we shall then prove that this cohomology class has the required property. The rest will be rather straightforward.

The component of dimension 2 of the single complex associated to $K_{\underline{e}}^{\cdot\cdot}$ has the form

$$(K_{\underline{e}}^{\cdot\cdot})^2 = K_{\underline{e}}^{0,2} \oplus K_{\underline{e}}^{1,1} \oplus K_{\underline{e}}^{2,0} = D^0(\underline{e}, C^2(-,\mathrm{Der}\_(-,0 \otimes \ker \Phi_{1,2})))$$

$$\oplus\ D^1(\underline{e}, C^1(-,\mathrm{Der}\_(-,0 \otimes \ker \Phi_{1,2}))) \oplus D^2(\underline{e}, C^0(-,\mathrm{Der}\_(-,0 \otimes \ker \Phi_{1,2})))$$

$$= \prod_{\substack{R \\ \pi\downarrow \\ A \\ \mu\downarrow \\ B}} \ \prod_{\substack{F_0 \xrightarrow{\alpha_1} F_1 \xrightarrow{\alpha_2} F_2 \\ \delta_0 \searrow \ \downarrow\delta_1 \ \swarrow \delta_2 \\ B}} \mathrm{Der}_A(F_0, B \otimes_A \ker \pi)$$

$$\oplus \prod_{\substack{R \longrightarrow R' \\ \pi\downarrow \quad \downarrow\pi' \\ A \longrightarrow A' \\ \mu\downarrow \quad \downarrow\mu' \\ B \longrightarrow B'}} \ \prod_{\substack{F_0 \xrightarrow{\alpha_1} F_1 \\ \delta_0 \searrow \ \swarrow \delta_1 \\ B}} \mathrm{Der}_A(F_0, B' \otimes_{A'} \ker \pi')$$

$$\oplus \prod_{\substack{R \longrightarrow R' \longrightarrow R'' \\ \pi\downarrow \quad \pi'\downarrow \quad \pi''\downarrow \\ A \longrightarrow A' \longrightarrow A'' \\ \mu\downarrow \quad \mu'\downarrow \quad \mu''\downarrow \\ B \longrightarrow B' \longrightarrow B''}} \ \prod_{\substack{F_0 \\ \delta_0\downarrow \\ B}} \mathrm{Der}_A(F_0, B'' \otimes_{A''} \ker \pi'') .$$

Let for every object $R \overset{\pi}{\to} A \overset{\mu}{\to} B$ of $\underline{e}$, $\sigma'_{\pi\mu}$ be a quasisection (see (1.2)) of the diagram

$$\begin{array}{ccc} - \underset{R}{\otimes} A : \underline{R\text{-free}} & \to & \underline{A\text{-free}}, \\ & \nwarrow^{\sigma'} & \uparrow \\ & & \underline{A\text{-free}}/_B \end{array}$$

and consider the 2-cochain 0 of $K_e^{\cdot\cdot}$ defined by

$$0_o(R \overset{\pi}{\to} A \overset{\mu}{\to} B)(\alpha_1,\alpha_2)) = (\sigma'_{\pi\mu}(\alpha_1\alpha_2) - \sigma'_{\pi\mu}(\alpha_1)\sigma'_{\pi\mu}(\alpha_2))(\delta_2 \underset{A}{\otimes} 1_{\ker \pi})$$

By construction 0_o is an element of the component $K_e^{o,2}$.

Let d_1 and d_2 denote the two differentials of the double complex $K_e^{\cdot\cdot}$. We already know (see (1.2)) that $d_2(0_o) = o$. Let us compute $d_1(0_o)$. We find

$$d_1(0_o) = -d_2(0_1)$$

where $0_1 \in K_e^{1,1}$ is given by

$$0_1 \begin{pmatrix} R & \overset{\beta_o}{\longrightarrow} & R' \\ \pi\downarrow & & \downarrow\pi' \\ A & \overset{\beta_1}{\longrightarrow} & A' \\ \mu\downarrow & & \downarrow\mu' \\ B & \overset{\beta_2}{\longrightarrow} & B' \end{pmatrix} \begin{pmatrix} F_0 & \overset{\alpha_1}{\longrightarrow} & F_1 \\ & \delta_o\searrow \quad \swarrow\delta_1 & \\ & B & \end{pmatrix} = (\beta_1 \underset{A}{\otimes} 1_{F_0})(\sigma'_{\pi'\mu'}(\alpha_1 \underset{A}{\otimes} 1_{A'}) -$$

$$\sigma'_{\pi\mu}(\alpha_1) \underset{R}{\otimes} 1_{R'})(\delta_1 \underset{A}{\otimes} 1_{\ker \pi'})\tau,$$

where τ is the morphism $B \underset{A}{\otimes} \ker \pi' \to B' \underset{A'}{\otimes} \ker \pi'$.

Moreover we observe that

$$d_1(0_1) = o.$$

Let $0 = 0_o + 0_1$ and let d be the differential of the simple complex associated to $K_e^{\cdot\cdot}$, then $d(0) = o$. Thus 0 defines a cohomology class $\underline{o} \in A^2(\underline{e},0)$.

Now $\underline{o} = 0$ is equivalent to the existence of an element $Q = Q_o + Q_1 \in (K''_e)^1 = K_e^{o,1} \oplus K_e^{1,o}$ such that

1. $o_o = d_2(-Q_o)$, 2. $o_1 = d_1(Q_o) - d_2(Q_1)$, 3. $0 = d_1(Q_1)$.

By the proof of (2.2.5) 1. is equivalent to the following statement: For all objects $(R \overset{\pi}{\to} A \overset{\mu}{\to} B)$ of $\underline{e}$ there exists a lifting $\sigma_o(B)$ of B as A-algebra to R, i.e. there exists a commutative diagram

$$\begin{array}{ccc} R & \longrightarrow & \sigma_o(B) \\ \pi\downarrow & & \downarrow\varepsilon \\ A & \longrightarrow & B \end{array}$$

such that $\sigma_o(B) \underset{R}{\otimes} A \simeq B$ and $\mathrm{Tor}_1^R(\sigma_o(B), A) = o$.

The set of such diagrams corresponds to the set of Q_o's with the property 1.. Given a Q_o with the property 1. then by the proof of (2.3.3) 2. is equivalent to the following statement: For every morphism $(\beta_o, \beta_1, \beta_2)$,

$$\begin{array}{ccc} R & \underset{\beta_o}{\longrightarrow} & R' \\ \pi\downarrow & & \downarrow\pi' \\ A & \underset{\beta_1}{\longrightarrow} & A' \\ \mu\downarrow & & \downarrow\mu' \\ B & \underset{\beta_2}{\longrightarrow} & B' \end{array}$$

of $\underline{e}$, there exists a morphism of rings $\sigma_o(\beta_2)$ making the following diagram commutative

$$\begin{array}{ccc} R & \underset{\beta_o}{\longrightarrow} & R' \\ \downarrow & & \downarrow \\ \sigma_o(B) & \underset{\sigma_o(\beta_2)}{\longrightarrow} & \sigma_o(B') \\ \varepsilon\downarrow & & \downarrow\varepsilon' \\ B & \underset{\beta_2}{\longrightarrow} & B' \end{array}$$

The set of such morphisms corresponds to the set of Q_1's with this property.

Finally 3. is equivalent to the following statement:

$$\sigma_o : \underline{e} \to 3.S\text{-}\underline{alg}$$

defined by

$$\sigma_o(R \to A \to B) = R \to \sigma_o(B) \to B$$

is a functor. This follows from an inspection of the proof of (2.3.3) and from (2.3.5).

The rest is straightforward. QED.

Corollary (4.1.15) Let $\underline{e}_0$ be any subcategory of $\underline{e}$.

Then the homomorphism

$$A^2(\underline{e},0) \to A^2(\underline{e}_0,0)$$

maps $o(\underline{e})$ onto $o(\underline{e}_0)$. Moreover, if both are zero, the map

$$\mathrm{Def}(\underline{e}) \to \mathrm{Def}(\underline{e}_0)$$

is a morphism of principal homogeneous spaces via the homomorphism

$$A^1(\underline{e},0) \to A^1(\underline{e}_0,0)\ .$$

Proof. This follows immediately from the proof of (4.1.14). QED.

Let $\underline{e}_0$ be any subcategory of the category of 3.S-algebras $\underline{e}$ (no conditions on $\underline{e}$ are needed), and suppose given a deformation σ_o of $\underline{e}_0$.

Definition (4.1.16) We shall denote by

$$\mathrm{Def}(\underline{e}/\underline{e}_0;\sigma_o)$$

the subset of $\mathrm{Def}(\underline{e})$ which maps to σ_o under the map $\mathrm{Def}(\underline{e}) \to \mathrm{Def}(\underline{e}_0)$.

Remark (4.1.17) Let σ be any deformation of $\underline{e}$, then the cochains $Q_o(\sigma(B))$ and $Q_1(\sigma(\beta_2))$ (denoted $Q_o(A')$ and $Q_1(\beta_2'))$ defined in the proofs of (2.2.5) and (2.3.3) respectively, fuse to define cochains $Q_o(\sigma)$ and $Q_1(\sigma)$ of $K^{\cdot\cdot}_{\underline{e}}$ characterizing the deformation σ.

Consider now a subcategory $\underline{e}_0$ of $\underline{e}$ and suppose we are given a deformation σ_o of $\underline{e}_0$. Then σ_o is characterized by the cochains $Q_o(\sigma_o)$ and $Q_1(\sigma_o)$ of $K^{o,1}_{\underline{e}_o}$ and $K^{1,o}_{\underline{e}_o}$ respectively. And we know that the obstruction cocycle $O(\underline{e}_o) = O_o(\underline{e}_0) + O_1(\underline{e}_0)$ of $K_{\underline{e}_0}$ is a coboundary, and in fact we have:

$$O_o(\underline{e}_0) = d_2(-Q_o(\sigma_o))$$

$$O_1(\underline{e}_0) = d_1(Q_o(\sigma_o)) - d_2(Q_1(\sigma_o))$$

$$0 = d_1(Q_1(\sigma_o)) .$$

Considering the short exact sequence of double complexes

$$o \to K^{\cdot\cdot}_{\underline{e}/\underline{e}_0} \to K^{\cdot\cdot}_{\underline{e}} \overset{\rho}{\to} K^{\cdot\cdot}_{\underline{e}_0} \to o$$

we find 1-cochains $Q_o' \in K^{o,1}_{\underline{e}}$ and $Q_1' \in K^{1,o}_{\underline{e}}$ such that $\rho(Q_o') = Q_o(\sigma_o)$, $\rho(Q_1') = Q_1(\sigma_o)$.

Let $Q' = Q_o' + Q_1'$.

Since $\rho(O(\underline{e}) - dQ') = o$ we find that

$$O(\underline{e}/\underline{e}_0) = O(\underline{e}) - dQ'$$

sits in $K^{\cdot\cdot}_{\underline{e}/\underline{e}_0}$. The corresponding cohomology class

$$o(\underline{e}/\underline{e}_0) \in A^2_{\underline{e}_0}(\underline{e},0)$$

depends only on the choice of σ_o .

Suppose there exists a deformation σ of $\underline{e}$ such that σ maps onto σ_o under the map

$$\mathrm{Def}(\underline{e}) \to \mathrm{Def}(\underline{e}_0)$$

then we have:

$$O_o(\underline{e}) = d_2(-Q_o(\sigma))$$

$$O_1(\underline{e}) = d_1(Q_o(\sigma)) - d_2(Q_1(\sigma))$$

$$o = d_1(Q_1(\sigma))$$

and, by construction, there exists a $\xi \in K^{o,o}_{\underline{e}_0}$, such that:

$$\rho(Q_o(\sigma)) - Q_o(\sigma_o) = d_2(\xi)$$

$$\rho(Q_1(\sigma)) - Q_1(\sigma_o) = d_1(\xi)$$

Pick a cochain $\xi' \in K^{o,o}_{\underline{e}}$ with $\rho(\xi') = \xi$, and put:

$$Q'_o = Q_o(\sigma) - d_1(\xi')$$

$$Q'_1 = Q_1(\sigma) - d_1(\xi')$$

Then $\rho(Q'_o) = Q_o(\sigma_o)$, $\rho(Q'_1) = Q_1(\sigma_o)$ and the corresponding

$$O(\underline{e}/\underline{e}_0) = O(\underline{e}) - dQ' = o ,$$

thus $o(\underline{e}/\underline{e}_0) = o$.

On the other hand suppose $o(\underline{e}/\underline{e}_0) = o$, then

$$O(\underline{e}/\underline{e}_0) = O(\underline{e}) - dQ' = dR$$

with $R = R_o + R_1 \in K^{\cdot\cdot}_{\underline{e}/\underline{e}_0}$.

In particular there exists a deformation σ of $\underline{e}$, and one with

$$Q_o(\sigma) = Q_o' + R_o \ , \quad Q_1(\sigma) = Q_1' + R_1$$

Since

$$\rho(Q_o(\sigma)) = \rho(Q_o'+R_o) = Q_o(\sigma_o)$$

$$\rho(Q_1(\sigma)) = \rho(Q_1'+R_1) = Q_1(\sigma_o)$$

we find that the map

$$\mathrm{Def}(\underline{e}) \to \mathrm{Def}(\underline{e}_0)$$

maps σ onto σ_o.

We have thus proved the follwoing result:

Theorem (4.1.17) Given a deformation σ_o of $\underline{e}_0$, then there is an obstruction

$$o(\underline{e}/\underline{e}_0;\sigma_o) \in A^2_{\underline{e}_o}(\underline{e},0)$$

such that $o(\underline{e}/\underline{e}_0;\sigma_o) = 0$ if and only if $\mathrm{Def}(\underline{e}/\underline{e}_0;\sigma_o)$ is non-empty.

In this case $\mathrm{Def}(\underline{e}/\underline{e}\ ;\sigma_o)$ is a principal homogeneous space over

$$A^1_{\underline{e}_o}(\underline{e},0)\ .$$

(4.2) Formal moduli

Let V be any local ring with maximal ideal $\underline{m}_V$ and residue field $k = V/\underline{m}_V$, and consider the category $\underline{l}$ of loca V-algebras of finite length with residue field k.

Given an object R of $\underline{l}$ we shall denote by $\underline{m}_R$ the maximal ideal of R. Thus $k = R/\underline{m}_R$.

There is a filtration of the category $\underline{l}$, the nth member of which is the full subcategory $\underline{l}_n$ of $\underline{l}$ defined by the objects R with $\underline{m}_R^n = 0$. Moreover there are functors

$$\lambda_n^{n+1} : \underline{l}_{n+1} \to \underline{l}_n \qquad n \geq 1$$

defined by

$$\lambda_n^{n+1}(R) = R/\underline{m}_R^n .$$

Consider any pair of subcategories $\underline{d}_0 \subseteq \underline{d}$ of 2.k-alg. We shall have to divide the further discussion into two cases.

Case 1. V is in this case supposed to be a k-algebra.

Case 2. V is in this case arbitrary, but we shall require $\underline{d}$, and therefore $\underline{d}_0$, to be a subcategory of k-alg, usually denoted $\underline{c}$ and $\underline{c}_0$ respectively.

Let R be any object of $\underline{l}$. We shall consider the following subcategories

$$\underline{e}_{0R} \subseteq \underline{e}_R$$

of 3.V-alg.

In case 1. the objects of $\underline{e}_R$ (resp. $\underline{e}_{0R}$) are the diagrams of the form

$$R \underset{k}{\otimes} A \to A \to B$$

where $(A \to B)$ is an object of $\underline{d}$ (resp. $\underline{d}_0$).
The morphisms of $\underline{e}_R$ (resp. $\underline{e}_{0R}$) are those induced by the morphisms of $\underline{d}$ (resp. $\underline{d}_0$).

In case 2. the objects of $\underline{e}_R$ (resp. $\underline{e}_{0R}$) are the diagrams of the form

$$R \to k \to B$$

where $(k \to B)$ is an object of $\underline{d}$ (resp. $\underline{d}_0$) (i.e. B is an object of $\underline{c}$ (resp. $\underline{c}_0$)). The morphisms are those induced by the morphisms of $\underline{d}$ (resp. $\underline{d}_0$).

With these notations, we shall define the functors

$$\mathrm{Def}(\underline{d}) : \underline{l} \to \underline{\mathrm{Sets}}$$
$$\mathrm{Def}(\underline{d}_o) : \underline{l} \to \underline{\mathrm{Sets}}$$

by:

$$\mathrm{Def}(\underline{d})(R) = \mathrm{Def}(\underline{e}_R)$$
$$\mathrm{Def}(\underline{d}_0)(R) = \mathrm{Def}(\underline{e}_{0R}) .$$

(We shall leave as an exercise the verifications needed to show that these objects are functors.)

In Case 1. both functors are pointed, in fact, R given, there is a canonical trivial deformation of $\underline{e}_R$ (resp. $\underline{e}_{0R}$) given by the diagrams

$$\begin{array}{ccc} R \underset{k}{\otimes} A & \longrightarrow & R \underset{k}{\otimes} B \\ \downarrow & & \downarrow \\ A & \longrightarrow & B \end{array} .$$

In Case 2. we shall assume that the functor $\mathrm{Def}(\underline{d}_0)$ is pointed

In both cases we shall denote the point of $\mathrm{Def}(\underline{d}_0)$ by $*$.

Let

$$\mathrm{Def}(\underline{d}/\underline{d}_o) : \underline{l} \to \underline{\mathrm{Sets}}$$

be the functor defined by

$$\mathrm{Def}(\underline{d}/\underline{d}_0)(R) = \mathrm{Def}(\underline{e}_R/\underline{e}_{0R}; *)$$

The purpose of this paragraph is to prove that this functor has a hull, and moreover, to give the structure of this hull.

We need some preparations. Let $\rho : R \to R'$ be any surjective morphism of $\underline{1}$ such that

$$\underline{m}_R \cdot \ker \rho = 0 .$$

Notice that in this case there are canonical isomorphisms of $\underline{1}$

$$R \underset{R'}{\times} R \;=\; R \underset{R'}{\times} R'[\ker \rho] \;=\; R \underset{k}{\times} k[\ker \rho]$$

producing the following commutative diagrams

$$\begin{array}{ccc} R \underset{R'}{\times} R & = & R \underset{k}{\times} k[\ker \rho] \\ \downarrow{\scriptstyle pr_1} & & \downarrow{\scriptstyle pr_1} \\ R & = & R \end{array} \qquad \begin{array}{ccc} R \underset{R'}{\times} R & = & R \underset{k}{\times} k[\ker \rho] \\ \downarrow{\scriptstyle pr_2} & & \downarrow{\scriptstyle \mu} \\ R & = & R \end{array}$$

where μ is defined by

$$\mu(r,(\alpha,x)) = r + x .$$

In this situation we shall prove the following souped up version of (4.1.17).

<u>Theorem (4.2.1)</u> Given an element $\bar{\sigma} \in \mathrm{Def}(\underline{d}/\underline{d}_o)(R')$ there is an obstruction

$$o(\bar{\sigma},\rho) \in A^2_{\underline{d}_o}(\underline{d}, O_{\underline{d}} \underset{V}{\otimes} \ker \rho)$$

such that $o(\bar{\sigma},\rho) = 0$ if and only if

$$\bar{\sigma} \in \operatorname{im} \mathrm{Def}(\underline{d}/\underline{d}_o)(\rho) .$$

In any case the diagrams above induce a commutative diagram

$$\begin{array}{ccc} \mathrm{Def}(\underline{d}/\underline{d}_o)(R \underset{R'}{\times} R) & = & \mathrm{Def}(\underline{d}/\underline{d}_o)(R \underset{R'}{\times} R'[\ker \rho]) \\ \downarrow{\scriptstyle p} & & \Vert \\ \mathrm{Def}(\underline{d}/\underline{d}_o)(R) \underset{\mathrm{Def}(\underline{d}/\underline{d}_o)(R')}{\times} \mathrm{Def}(\underline{d}/\underline{d}_o)(R) & \overset{\mu^1}{\leftarrow} & \mathrm{Def}(\underline{d}/\underline{d}_o)(R) \times A^1_{\underline{d}_o}(\underline{d}, O_{\underline{d}} \underset{V}{\otimes} \ker \rho) \\ \downarrow{\scriptstyle pr_1} & & \downarrow{\scriptstyle pr_1} \\ \mathrm{Def}(\underline{d}/\underline{d}_o)(R) & = & \mathrm{Def}(\underline{d}/\underline{d}_o)(R) \end{array}$$

in which the maps p and μ^1 are surjections.

Proof. Let $\bar{\sigma} \in \mathrm{Def}(d/d_0)(R')$ and pick a representative σ of the equivalence class $\bar{\sigma}$. Consider the following subcategory e of 3.V-alg.

In Case 1. the objects of e are the diagrams

$$R \underset{k}{\otimes} A \to R' \underset{k}{\otimes} A \to \sigma(R' \underset{k}{\otimes} A \to A \to B)$$

where $A \to B$ is an object of $\underline{d}$. The morphisms of $\underline{e}$ are those induced by the morphisms of $\underline{d}$. Obviously $\underline{d}_0$ corresponds to a subcategory $\underline{e}_0$ of $\underline{e}$.

In Case 2. the objects of $\underline{e}$ are the diagrams

$$R \to R' \to \sigma(R' \to S \to B)$$

where $S \to B$ is an object of $\underline{d}$ (i.e. B is an object of $\underline{c}$). The morphisms of $\underline{e}$ are those induced by the morphisms of $\underline{d}$ (i.e. $\underline{c}$). Obviously $\underline{d}_0$ corresponds to a subcategory $\underline{e}_0$ of $\underline{e}$.

By (4.1.17) there is an obstruction

$$o(\underline{e}/\underline{e}_0;*) \in A^2_{\underline{e}_o}(\underline{e},0)$$

such that $o(\underline{e}/\underline{e}_0;*) = 0$ if and only if there is a deformation of $\underline{e}$ reducing to * on $\underline{e}_0$. The first part of the theorem then follows from the existence of canonical isomorphisms:

$$A^n_{\underline{e}_0}(\underline{e},0) = A^n_{\underline{d}_o}(\underline{d},0_{\underline{d}} \underset{k}{\otimes} \ker \rho), \qquad n \geq 0$$

The cohomology on the left side is given by the double complex

$$K^{\cdot\cdot}_{\underline{e}/\underline{e}_0} = D^{\cdot}_{\underline{e}_0}(\underline{e},C^{\cdot}(-,\mathrm{Der}_-(-,0 \otimes \ker \Phi_{1,2})))$$

In Case 1. let

$$\begin{array}{ccc}
R \underset{k}{\otimes} A_1 & \overset{\beta_0}{\to} & R \underset{k}{\otimes} A_2 \\
\downarrow \pi_1 & & \downarrow \pi_2 \\
R' \underset{k}{\otimes} A_1 & \overset{\beta_1}{\to} & R' \underset{k}{\otimes} A_2 \\
\downarrow & & \downarrow \\
\sigma(R' \underset{k}{\otimes} A_1 \to A_1 \to B_1) & \overset{\beta_2}{\to} & \sigma(R' \underset{k}{\otimes} A_2 \to A_2 \to B_2)
\end{array}$$

be an object of Mor $\underline{e}$, then

$$C^{\cdot}(-,\mathrm{Der}_{-}(-,0 \otimes \ker \Phi_{1,2}))(\beta_0,\beta_1,\beta_2) =$$

$$C^{\cdot}((R' \otimes A_1)\text{-}\underline{\mathrm{free}}/\sigma(R' \otimes A_1 \to A_1 \to B_1)^{o}, \mathrm{Der}_{R'\otimes A_1}(-,\sigma(R' \otimes A_2 \to A_2 \to B_2) \underset{\substack{R'\otimes A_2 \\ k}}{\otimes} \ker \pi_2))$$

Now $\ker \pi_2 = A_2 \underset{k}{\otimes} \ker \rho$, therefore

$$\sigma(R' \otimes A_2 \to A_2 \to B_2) \underset{R'\otimes A_2}{\otimes} \ker \pi_2 = B_2 \otimes \ker \rho .$$

There exists a canonical functor

$$(R' \otimes A_1)\text{-}\underline{\mathrm{free}}/\sigma(R' \otimes A_1 \to A_1 \to B_1) \to A_1\text{-}\underline{\mathrm{free}}/B_1$$

defined by tensorization with A_1 over $R' \otimes A_1$. This functor induces a morphism of complexes:

$$\begin{array}{l}
C^{\cdot}(A_1\text{-}\underline{\mathrm{free}}/B_1^{o}, \mathrm{Der}_{A_1}(-,B_2 \otimes \ker \rho)) \\
\downarrow \\
C^{\cdot}(-,\mathrm{Der}_{-}(-,0 \otimes \ker \emptyset_{1,2}))(\beta_0,\beta_1,\beta_2)
\end{array}$$

Notice that $\underline{e}$ (resp. $\underline{e}_0$) is, in a natural way, isomorphic to $\underline{d}$ (resp. $\underline{d}_0$). Thus the morphism above induces a morphism of double complexes

$$K^{\cdot\cdot}_{\underline{d}/\underline{d}_0}(0_{\underline{d}} \otimes \ker \rho) \to K^{\cdot\cdot}_{\underline{e}/\underline{e}_0} .$$

Due to a result of André (see (An) (19.2) p. 78) the corresponding morphism of the first spectral sequences is an isomorphism. This ends the proof of the first part of the theorem.

The only remaining difficulty is the following. Suppose $(\bar\sigma_1,\bar\sigma_2)$ is an element of

$$\mathrm{Def}(\underline{d}/\underline{d}_0)(R) \underset{\mathrm{Def}(\underline{d}/\underline{d}_0)(R')}{\times} \mathrm{Def}(\underline{d}/\underline{d}_0)(R)$$

then by definition the map

$$\mathrm{Def}(\underline{d}/\underline{d}_0)(\rho): \mathrm{Def}(\underline{d}/\underline{d}_0)(R) \to \mathrm{Def}(\underline{d}/\underline{d}_0)(R')$$

maps $\bar\sigma_1$ and $\bar\sigma_2$ onto the same element. But remember that we are talking about classes of deformations. This implies that if σ_1 and σ_2' are representatives of $\bar\sigma_1$ and $\bar\sigma_2$ respectively, then $\sigma_1 \underset{R}{\otimes} R'$ and $\sigma_2' \underset{R}{\otimes} R'$ are equivalent deformations of $\underline{e}_{R'}$, but they need not be equal.
However, if

$$\theta' : \sigma_1 \underset{R}{\otimes} R' \xrightarrow{\sim} \sigma_2' \underset{R}{\otimes} R'$$

is an equivalence, we easily prove (see the Appendix) that there is a third deformation σ_2 of $\underline{e}_R$ and an equivalence

$$\theta : \sigma_2 \xrightarrow{\sim} \sigma_2'$$

such that:

$$\theta \underset{R}{\otimes} R' = \theta'$$

Therefore σ_2 is another representative of $\bar\sigma_2$ and this time we have

$$\sigma_1 \underset{R}{\otimes} R' = \sigma_2 \underset{R}{\otimes} R'$$

With this done, we shall complete the proof of (4.2.1) by establishing the surjectivity of p.
Keeping the notations above we have to construct a deformation σ_o of $\underline{e}_{R \underset{R'}{\times} R}$ such that

$$\mathrm{Def}(\underline{d}/\underline{d}_0)(pr_i)(\sigma_o) = \sigma_i \qquad i = 1,2 .$$

This construction is, in full generality, both lengthy and dull, though completely elementary. The point will be equally well understood restricting our situation to the following simple one:

$$\underline{d} = \{S \to A\} \qquad \underline{d}_o = \emptyset$$
$$R' = S .$$

Corresponding to the liftings σ_1 and σ_2 there is a commutative diagram

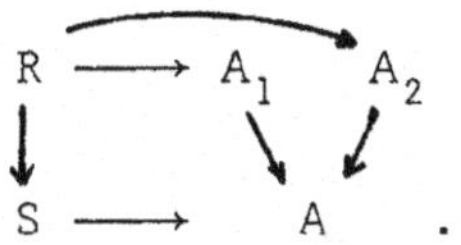

We know that

$$A_i = \varinjlim_{S\text{-}\underline{free}/A} \sigma'(A_i) \qquad i = 1,2$$

(see the proof of (2.2.5)) where $\sigma'(A_i)$ are the quasisection corresponding to the liftings A_i $i = 1,2$.

Let for any ring T, $T[X]$ denote the polynomial algebra on one variable. Then since

$$R[X] \underset{S[X]}{\times} R[X] = R \underset{S}{\times} R[X]$$

the two f-quasisections $\sigma'(A_1)$ and $\sigma'(A_2)$ fuse to give an f-quasisection

$$\sigma'(A_1) \underset{f}{\times} \sigma'(A_2)$$

of the functor

$$(R \underset{S}{\times} R)\text{-}\underline{free} \to S\text{-}\underline{free} .$$

Since the obstruction cocycles of $\sigma'(A_1)$ and $\sigma'(A_2)$ both are zero, the obstruction cocycle of $\sigma'(A_1) \underset{f}{\times} \sigma'(A_2)$ is also zero, therefore

$$A_o = \underset{\substack{\rightarrow \\ \underline{S\text{-free}}/A}}{\lim} \sigma'(A_1) \underset{f}{\times} \sigma'(A_2)$$

is a lifting of A to $R \underset{S}{\times} R$. which, by construction has the properties required.

Moreover, via the canonical isomorphisms

$$\mathrm{Def}(\underline{d}/\underline{d}_o)(R \underset{S}{\times} R) \simeq \mathrm{Def}(\underline{d}/\underline{d}_o)(R) \times \mathrm{Def}(\underline{d}/\underline{d}_o)(S[\ker \rho])$$

$$\mathrm{Def}(\underline{d}/\underline{d}_o)(R \underset{S}{\times} R) \simeq \mathrm{Def}(\underline{d}/\underline{d}_o)(R) \times A^1_{\underline{d}_o}(\underline{d}, O_{\underline{d}} \otimes \ker \rho)$$

induced by the canonical isomorphisms

$$R \underset{S}{\times} R \simeq R \underset{S}{\times} S[\ker \rho],$$

this lifting A_o, corresponds to the pairs (A_1, A_{21}) and $(A_1, \lambda(A_2, A_1))$ respectively, where

$$A_{21} = \underset{\substack{\rightarrow \\ \underline{S\text{-free}}/A}}{\lim} \sigma_{21}$$

with the f-quasisection σ_{21} of

$$S[\ker\rho]\text{-}\underline{\text{free}} \rightarrow S\text{-}\underline{\text{free}}$$

defined by

$$\sigma_{21}\begin{pmatrix} F_o & \xrightarrow{\psi} & F_1 \\ & \searrow \quad \swarrow & \\ & A & \end{pmatrix} = F_o \underset{S}{\otimes} S[\ker \rho] \xrightarrow{\sigma_{21}(\psi)} F_1 \underset{S}{\otimes} S[\ker \rho]$$

where for $x \in F_o$,

$$\sigma_{21}(\psi)(x) = \psi(x) + (\sigma'(A_2)(\psi) - \sigma'(A_1)(\psi))(x)$$

which is meaningfull since all coefficients of the polynomial

$$\sigma'(A_2)(\psi)(x) - \sigma'(A_1)(\psi)(x)$$

sit in $\ker\rho$.

This ends the proof of (4.2.1). QED.

<u>Remark (4.2.2)</u> We have tacitely used the fact that a deformation of a deformation is a deformation. This follows from elementary diagram chasing

<u>Corollary (4.2.3)</u> Consider a commutative diagram of $\underline{l}$ of the form

$$\begin{array}{ccc} R_1 & \xrightarrow{\rho} & R_2 \\ {\scriptstyle \rho_1}\downarrow & & \downarrow{\scriptstyle \rho_2} \\ R_1' & \xrightarrow{\rho'} & R_2' \end{array}$$

and suppose

$$\underline{m}_{R_1} \cdot \ker\rho_1 = \underline{m}_{R_2} \cdot \ker\rho_2 = 0 .$$

Let $\bar{\sigma}_1$ be an element of $\mathrm{Def}(\underline{d}/\underline{d}_0)(R_1')$ and put $\bar{\sigma}_2 = \mathrm{Def}(\underline{d}/\underline{d}_0)(\rho')(\bar{\sigma}_1)$.

Consider the homomorphisms induced by ρ ,

$$\rho_*^n : A^n_{\underline{d}_o}(\underline{d}, 0_{\underline{d}} \otimes \ker\rho_1) \to A^n_{\underline{d}_o}(\underline{d}, 0_{\underline{d}} \otimes \ker\rho_2) .$$

Then

$$\rho_*^2(o(\bar{\sigma}_1, \rho_1)) = o(\bar{\sigma}_2, \rho_2) .$$

Moreover the induced diagram

$$\begin{array}{ccc} \mathrm{Def}(\underline{d}/\underline{d}_0)(R_1) \times A^1_{\underline{d}_o}(\underline{d}, 0_{\underline{d}} \otimes \ker\rho_1) & \to & \mathrm{Def}(\underline{d}/\underline{d}_0)(R_2) \times A^1_{\underline{d}_o}(\underline{d}, 0_{\underline{d}} \otimes \ker\rho_2) \\ \downarrow{\scriptstyle \mu^1} & & \downarrow{\scriptstyle \mu^1} \\ \mathrm{Def}(\underline{d}/\underline{d}_0)(R_1) \underset{\mathrm{Def}(\underline{d}/\underline{d}_0)(R_1')}{\times} \mathrm{Def}(\underline{d}/\underline{d}_0)(R_1) & \to & \mathrm{Def}(\underline{d}/\underline{d}_0)(R_2) \underset{\mathrm{Def}(\underline{d}/\underline{d}_0)(R_2')}{\times} \mathrm{Def}(\underline{d}/\underline{d}_0)(R_2) \end{array}$$

commutes.

Proof. This follows immediately from the definitions. QED.

Consider the category of k-vector spaces, k-mod. Let W be any object of k-mod. Pick a basis $\{v_i\}_{i\in I}$ for W and put the topology on W in which a basis for the neighbourhoods of the neutral element consits of the subspaces containing all but a finite number of the elements v_i. Consider the corresponding category of topological k-vector spaces, k-top.mod. Let Hom_k^c denote the Hom functor in this category. Obviously all finite dimensional vector spaces will be discrete. Moreover there is a natural topology, defined by the dual basis, on the topological dual $W^* = \mathrm{Hom}_k^c(W,k)$ of any object W of k-top.mod. And one easily checks that there is a canonical isomorphism

$$W \simeq W^{**}.$$

We shall now use these generalities in the construction of a hull for $\underline{\mathrm{Def}}(\underline{d}/\underline{d}_o)$.

Fix a basis $\{x_i\}_{i\in I}$ for $A^1_{\underline{d}_o}(\underline{d},O_{\underline{d}})$ and a basis $\{y_j\}_{j\in J}$ for $A^2_{\underline{d}_o}(\underline{d},O_{\underline{d}})$, and consider the corresponding topological duals $A^i_{\underline{d}_o}(\underline{d},O_{\underline{d}})^*$. Denote by

$$\mathrm{Free}_V(A^i_{\underline{d}_o}(\underline{d},O_{\underline{d}})^*) \qquad i = 1,2,$$

the polynomial V-algebra on the (dual) basis of $A^i_{\underline{d}_o}(\underline{d},O_{\underline{d}})^*$. Thus $\mathrm{Free}_V(A^i_{\underline{d}_o}(\underline{d},O_{\underline{d}})^*) \underset{V}{\otimes} k \simeq \mathrm{Sym}_k(A^i_{\underline{d}_o}(\underline{d},O_{\underline{d}})^*)$.

Consider the topology of $\mathrm{Free}_V(A^i_{\underline{d}_o}(\underline{d},O_{\underline{d}})^*)$ defined by the ideals generated by an open subset of $A^i_{\underline{d}_o}(\underline{d},O_{\underline{d}})^*$ together with some power of the obvious maximal ideal.

Let

$$T^i, \quad i = 1,2$$

be the completion of $\mathrm{Free}_V(A^i_{\underline{d}_o}(\underline{d},0_{\underline{d}_o})^*)$ in this topology. Then T^i is a profinite local ring. In particular, if $A^i_{\underline{d}_o}(\underline{d},0_{\underline{d}})$ has finite k-dimension, T^i is the V-algebra of convergent powerseries in the (dual) basis of $A^i_{\underline{d}_o}(\underline{d},0_{\underline{d}})^*$. Moreover, if $A^i_{\underline{d}_o}(\underline{d},0_{\underline{d}})$ has a countable basis, then in the corresponding topology on T^i there is a countable basis for the system of neighbourhoods of the neutral element.

<u>Theorem (4.2.4)</u> Suppose $A^1_{\underline{d}_o}(\underline{d},0_{\underline{d}})$ has a countable basis as a k-vector space. Pick such a basis $\{x_i\}_{i\in\mathbb{N}}$, then there exists a morphism of complete local rings

$$o : T^2 \to T^1$$

such that

$$H(\underline{d}/\underline{d}_o) = T^1 \hat{\otimes}_{T^2} V$$

is a hull for the functor $\mathrm{Def}(\underline{d}/\underline{d}_o)$.

<u>Proof</u>. For each $n \in \mathbb{N}$ let's put $T^i_n = T^i/\underline{m}_{T^i}{}^n$. T^i_n has a natural topology, the quotient of the topology of T^i. Our first step will be to prove that in Case 1. T^1_2 prorepresent the functor $\mathrm{Def}(\underline{d}/\underline{d}_o)$ restricted to $\underline{1}_2$. This is rather easy. In fact let R be any object of $\underline{1}_2$. By definition $\underline{m}_R^2 = 0$ implying that R as a commutative ring (k-algebra) is equal to $k[\underline{m}_R]$, the Nagata ring of the k-vectorspace $\underline{m}_R$. Now let Mor^c_V denote the set of continuous morphisms in the category of topological local rings and consider the canonical isomorphisms:

$$\mathrm{Mor}^c_V(T^1_2,R) = \mathrm{Mor}^c_V((V/\underline{m}_V^2)[A^1_{\underline{d}_o}(\underline{d},0_{\underline{d}})^*],R)$$

$$= \mathrm{Hom}^c_k(A^1_{\underline{d}_o}(\underline{d},0_{\underline{d}})^*,\underline{m}_R) = A^1_{\underline{d}_o}(\underline{d},0_{\underline{d}}) \otimes_k \underline{m}_R = A^1_{\underline{d}_o}(\underline{d},0_{\underline{d}} \otimes \underline{m}_R).$$

Since in Case 1. every object R of $\underline{1}$ is a k-algebra and all morphisms are k-algebra morphisms there is a canonical element σ_o in $\mathrm{Def}(\underline{d}/\underline{d}_0)(R)$ namely the trivial deformation $* = \mathrm{Def}(\underline{d}/\underline{d}) \subseteq \mathrm{Def}(\underline{d}/\underline{d}_0)$.

By (4.2.1) this implies that there is a canonical isomorphism

$$A^1_{\underline{d}_0}(\underline{d}, O_{\underline{d}} \otimes \underline{m}_R) \simeq \mathrm{Def}(\underline{d}/\underline{d}_0)(R)$$

proving what we wanted to prove.

In Case 2. there is, as in Case 1., an initial object of $\underline{1}_2$, namely $V/\underline{m}_V^2$ and by (4.2.1) there is an obstruction

$$o_1 \in A^2(\underline{d}, O_{\underline{d}} \underset{k}{\otimes} \underline{m}_V^2)$$

which is zero if and only if $\mathrm{Def}(\underline{d})(V/\underline{m}_V^2)$ is non-empty.

Now, we have canonical isomorphisms and a canonical inclusion

$$A^2(\underline{d}, O_{\underline{d}} \underset{k}{\otimes} \underline{m}_V/\underline{m}_V^2) = \mathrm{Hom}_V^C(A^2(\underline{d}, O_{\underline{d}})^*, \underline{m}_V/\underline{m}_V^2)$$

$$= \mathrm{Mor}_V^C((V/\underline{m}_V^2)[A^2(\underline{d}, O_{\underline{d}})^*], V/\underline{m}_V^2) = \mathrm{Mor}_V^C(T_2^2, V/\underline{m}_V^2) \subseteq \mathrm{Mor}_V^C(T_2^2, T_2^1).$$

Let R be any object of $\underline{1}_2$ and consider the canonical morphism

$$\nu : V/\underline{m}_V^2 \to R.$$

We know (see (4.2.3)) that the image of o_1, considered as an element of $\mathrm{Mor}_V^C(T_2^2, V/\underline{m}_V^2)$, in $\mathrm{Mor}_V^C(T_2^2, R) = A^2(\underline{d}, O_{\underline{d}} \otimes \underline{m}_R)$ under the map induced by ν is zero if and only if $\mathrm{Def}(\underline{d})(R)$ is non-empty. Let

$$V_2 = V/\underline{m}_V^2 \underset{T_2^2}{\otimes} V/\underline{m}_V^2, \quad H_2 = T_2^1 \underset{T_2^2}{\otimes} V/\underline{m}_V^2 = V_2[A^1(\underline{d}, O_{\underline{d}})^*]$$

$V/\underline{m}_V^2$ and T_2^1 being considered as T_2^2-modules via the morphism

$o_1 \in \mathrm{Mor}_V^c(T_2^2, V/\underline{m}_V{}^2) \subseteq \mathrm{Mor}_V^c(T_2^2, T_2^1)$. V_2 is the largest quotient of $V/\underline{m}_V{}^2$ to which $\underline{d}$ may be lifted.

Since we know that $\mathrm{Def}(\underline{d})(V_2)$ is non-empty, we may pick an element σ_1 in $\mathrm{Def}(\underline{d})(V_2)$. This element will take the place of the trivial deformation in Case 1.

For any object R of $\underline{l}_2$ we have canonical isomorphisms

$$\mathrm{Mor}_V^c(H_2, R) = \mathrm{Mor}_V^c(V_2[A^1(\underline{d}, O_{\underline{d}})^*], R)$$

$$= \begin{cases} \emptyset \text{ if } \nu : V/\underline{m}_V{}^2 \to R \text{ does not factor through } V_2 \\ \mathrm{Hom}_k^c(A^1(\underline{d}, O_{\underline{d}})^*, \underline{m}_R) = A^1(\underline{d}, O_{\underline{d}} \otimes \underline{m}_R) \text{ if } \nu \text{ does factor through } V_2 . \end{cases}$$

Using the element $\sigma_1 \in \mathrm{Def}(\underline{d})(V_2)$ as neutral element we find functorial isomorphisms

$$A^1(\underline{d}, O_{\underline{d}} \otimes \underline{m}_R) \simeq \mathrm{Def}(\underline{d})(R)$$

whenever the latter is non-empty, thus proving the existence of a natural isomorphism of functors,

$$\mathrm{Mor}_V^c(H_2, -) \simeq \mathrm{Def}(\underline{d})$$

on the category $\underline{l}_2$.

Let in Case 1. $o_1 : T_2^2 \to T_2^1$ be the trivial morphism (i.e. the composition $T_2^2 \to V/\underline{m}^2 \to T_2^1$). We have then, in both cases, proved the following statement:

There exists a continuous morphism

$$o_1 : T_2^2 \to T_2^1$$

such that the corresponding closed fiber

$$H_2 = T_2^1 \underset{T_2^2}{\otimes} V/\underline{m}_V{}^2$$

prorepresents the functor $\mathrm{Def}(\underline{d}/\underline{d}_o)$ restricted to $\underline{l}_2$.

In order to extend this result to all subcategories $\underline{1}_n$ of $\underline{1}$ we shall have to take a closer look at the isomorphism

$$\mathrm{Mor}_V^c(H_2,-) \simeq \mathrm{Def}(\underline{d}/\underline{d}_o) .$$

This can be done in the following way. Let I_2 denote the set of open ideals of H_2 and consider the following subcategory $\underline{e}_2$ of 3.V-$\underline{\mathrm{Alg}}$. An object of $\underline{e}_2$ is, <u>in Case 1</u>. a diagram

$$A \underset{k}{\otimes} H_2/\mathfrak{a} \to A \to B$$

where $A \to B$ is an object of $\underline{d}$ and $\mathfrak{a} \in I_2$. The morphisms of $\underline{e}_2$ are those induced by the morphisms of $\underline{d}$ and by the morphisms of the form $H_2/\mathfrak{a} \to H_2/\mathfrak{b}$ with $\mathfrak{a},\mathfrak{b} \in I_2$, $\mathfrak{a} \subseteq \mathfrak{b}$. Obviously the subcategory $\underline{d}_o$ of $\underline{d}$ corresponds to a subcategory $\underline{e}_{o2}$ of $\underline{e}_2$.

<u>In Case 2</u>. the objects of $\underline{e}_2$ are the diagrams

$$H_2/\mathfrak{a} \to k \to B$$

where $k \to B$ is an object of $\underline{d}$ (i.e. B is an object of $\underline{c}$) and $\mathfrak{a} \in I_2$. The morphisms of $\underline{e}_2$ being those induced by the morphisms of $\underline{d}$ (i.e. $\underline{c}$) and the morphisms

$$H_2/\mathfrak{a} \to H_2/\mathfrak{b}$$

above. Obviously $\underline{d}_o$ corresponds to a subcategory $\underline{e}_{o2}$ of $\underline{e}_2$

If we consider I_2 as an ordered set (by inclusion), therefore as a category, we find

$$\underline{e}_2 \simeq \underline{d} \times I_2 , \quad \underline{e}_{o2} \simeq \underline{d}_o \times I_2 .$$

Now we apply (4.2.1). There is an obstruction

$$\underline{o} \in A^2_{\underline{e}_{o2}}(\underline{e}_2, O)$$

such that $\underline{o} = 0$ if and only if $\underline{e}_2$ admits a deformation trivial on $\underline{e}_{02}$. In that case the set of such deformations modulo isomorphisms is a principal homogenous space over $A^1_{\underline{e}_{02}}(\underline{e}_2, 0)$. Using ((La 1)(5.3)) and the isomorphisms $\underline{e}_2 \simeq \underline{d} \times I_2$, $\underline{e}_{02} \simeq \underline{d}_o \times I_2$, we find a spectral sequence given by the term

$$E_2^{p,q} = \varprojlim_{I_2}{}^{(p)} A^q_{\underline{d}_o}(\underline{d}, O_{\underline{d}}) \otimes_k (\underline{m}_{H_2}/\mathfrak{a})$$

converging to $A^{p+q}_{\underline{e}_{02}}(\underline{e}_2, 0)$.

Since I_2 contains a countable cofinal subset, and since the morphisms in the projective system above are all epimorphisms, $E_2^{p,q} = 0$ for $p \neq 0$.

Consequently

$$A^1_{\underline{e}_{02}}(\underline{e}_2, 0) = \varprojlim_{I_2} A^1_{\underline{d}_o}(\underline{d}, O_{\underline{d}}) \otimes_k (\underline{m}_{H_2}/\mathfrak{a})$$

$$A^2_{\underline{e}_{02}}(\underline{e}_2, 0) = \varprojlim_{I_2} A^2_{\underline{d}_o}(\underline{d}, O_{\underline{d}}) \otimes_k (\underline{m}_{H_2}/\mathfrak{a}).$$

Notice that for each $\mathfrak{a} \in I_2$ the category $\underline{e}_{(H_2/\mathfrak{a})}$ (see above) is a subcategory of $\underline{e}_2$. The obstruction for deforming $\underline{e}_{(H_2/\mathfrak{a})}$ relative to $\underline{e}_{o(H_2/\mathfrak{a})}$ sits in $A^2_{\underline{d}_o}(\underline{d}, 0) \otimes (\underline{m}_{H_2}/\mathfrak{a})$.

We already know that this obstruction is zero. Using this we find $\underline{o} = 0$.

Moreover, there is a nice canonical element of $A^1_{\underline{e}_{02}}(\underline{e}_2, 0)$. In fact, we observe that

$$A^1_{\underline{e}_{02}}(\underline{e}_2, 0) = \varprojlim_{\mathfrak{a} \in I_2} A^1_{\underline{d}_o}(\underline{d}, O_{\underline{d}}) \otimes (\underline{m}_{V_2} \oplus A^1_{\underline{d}_o}(\underline{d}, O_{\underline{d}})^*)/\mathfrak{a}).$$

The converging sum

$$\sigma_2 = \sum_k x_k \otimes \overline{(0,x_k^*)}$$

where $\overline{(0,x_k^*)}$ is the image in $(\underline{m}_{V_2} \oplus A_{\underline{d}_o}(\underline{d},0_{\underline{d}})^*)/\mathfrak{a}$ of the element $(0,x_k^*)$ of $\underline{m}_{V_2} \oplus A^1_{\underline{d}_o}(\underline{d},0_{\underline{d}})^*$, then defines an element of $A^1_{\underline{e}_{o2}}(\underline{e}_2,0)$.

Since we may, exactly as above, identify $A^1_{\underline{e}_{o2}}(\underline{e}_2,0)$ with the set of isomorphism classes of deformations of $\underline{e}_2$ relative to $\underline{e}_{o2}$, σ_2 corresponds to an isomorphism class of deformations of $\underline{e}_2$ relative to $\underline{e}_{o2}$. We shall pick a representative of this class and, abusing the language, we shall let σ_2 denote this representative.

Thus we find an element

$$\sigma_2 \in \varprojlim_{\underline{I}_2} \mathrm{Def}(\underline{d}/\underline{d}_o)(H_2/\mathfrak{a})$$

and we may convince ourselves about the fact that σ_2 determines the isomorphism of functors on $\underline{l}_2$:

$$\psi_2 : \mathrm{Mor}^c(H_2,-) \simeq \mathrm{Def}(\underline{d}/\underline{d}_o) .$$

Now let H_m for any m be a topological quotient of T^1_m and let I_m denote the ordered set of open ideals of H_m.

Let

$$\underline{e}_m \quad (\text{resp. } \underline{e}_{om}) \qquad\qquad m \leq n$$

denote the following subcategory of 3.V-alg. An object of $\underline{e}_m$ (resp. $\underline{e}_{om}$) is in Case 1. a diagram of the form

$$(H_m/\mathfrak{a}) \otimes_k A \to A \to B$$

where $A \to B$ is an object of $\underline{d}$ (resp. $\underline{d}_o$) and $\mathfrak{a} \in I_m$. The morphisms of $\underline{e}_m$ (resp. $\underline{e}_{om}$) are those induced by the morphisms of $\underline{d}$ (resp. $\underline{d}_o$) and by inclusions among the $\mathfrak{a}$'s.

In Case 2. the objects of $\underline{e}_m$ (resp. $\underline{e}_{om}$) are the diagrams of the form

$$H_m/\mathfrak{a} \to k \to B$$

where $k \to B$ is an object of $\underline{d}$ (resp. $\underline{d}_o$). The morphisms are those induced by the morphisms of $\underline{d}$ (resp. $\underline{d}_o$) and by the inclusions among the $\mathfrak{a}$'s .

By induction on m we shall prove that there exists for every m a topological quotient H_m of T^1_m and a deformation σ_m of $\underline{e}_m$ such that the restriction of σ_m to $\underline{e}_{om}$ is * , satisfying the following conditions:

1) The canonical morphism of topological rings

$$t_m : T^1_m \to T^1_{m-1}$$

induces a morphism of topological rings

$$h_m : H_m \to H_{m-1}$$

2) The corresponding map

$$\varprojlim_{I_m} \mathrm{Def}(\underline{d}/\underline{d}_o)(H_m/\mathfrak{a}) \to \varprojlim_{I_{m-1}} \mathrm{Def}(\underline{d}/\underline{d}_o)(H_{m-1}/\mathfrak{a})$$

maps σ_m onto σ_{m-1} .

3) σ_m determines a smooth morphism of functors on $\underline{l}_m$

$$\psi_m : \mathrm{Mor}^c(H_m,-) \to \mathrm{Def}(\underline{d}/\underline{d}_o)$$

inducing an isomorphism on the tangent spaces.

Assume such H_m and σ_m exist for $m \leq n$, and put

$$C_m = \ker(T^1_m \to H_m) , \qquad m \leq n ,$$

$$C_m' = \underline{m}_{T^1_{m+1}} \cdot \ker(T^1_{m+1} \to H_m) , \qquad m \leq n ,$$

$$H_m' = T^1_{m+1}/C_m' , \qquad m \leq n .$$

Let

$$h_{m+1}' : H_m' \to H_m , \qquad m \leq n$$

be the canonical morphism of topological rings. Obviously

$$\underline{m}_{H_m}' \cdot \ker h_{m+1}' = 0 , \qquad m \leq n .$$

Let I_m' be the ordered set of open ideals of H_m', and let

$$\underline{e}_n' \quad (\text{resp. } \underline{e}_{on}')$$

be the following subcategory of 3.V-$\underline{alg}$. In Case 1. the objects of $\underline{e}_n'$ (resp. $\underline{e}_{on}'$) are the diagrams of the form

$$(H_n'/\mathfrak{a}) \underset{k}{\otimes} A \to (H_n/h_{n+1}'(\mathfrak{a})) \underset{k}{\otimes} A \to \sigma_n((H_n/h_{n+1}'(\mathfrak{a})) \underset{k}{\otimes} A \to A \to B)$$

with $\mathfrak{a} \in I_n'$ and $A \to B$ an object of $\underline{d}$ (resp. $\underline{d}_o$). In Case 2. the objects are the diagrams of the form

$$(H_n'/\mathfrak{a}) \to H_n/h_{n+1}'(\mathfrak{a}) \to \sigma_n(H/h_{n+1}'(\mathfrak{a}) \to k \to B)$$

where $\mathfrak{a} \in I_n'$, and $k \to B$ is an object of $\underline{d}$ (resp. $\underline{d}_o$). The morphisms being defined accordingly.

We observe that

$$\underline{e}_n' \simeq \underline{d} \times I_n' \quad (\text{resp. } \underline{e}_{on}' \simeq \underline{d}_o \times I_n') .$$

The obstruction for deforming $\underline{e}_n'$ relative to $\underline{e}_{on}'$, i.e. such that the restriction to $\underline{e}_{on}'$ is $*$, is an element

$$\underline{o}_n \in A^2_{\underline{e}_{on}'}(\underline{e}_n', 0) = \varprojlim_{I_n'} A^2_{\underline{d}_o}(\underline{d}, O_{\underline{d}}) \otimes ((\ker h_{n+1}' + \mathfrak{a})/\mathfrak{a})$$

$$= \varprojlim_{I_n'} \mathrm{Hom}_k^c(A^2_{\underline{d}_o}(\underline{d}, O_{\underline{d}})^*, ((\ker h_{n+1}' + \mathfrak{a})/\mathfrak{a})) .$$

Given any $\mathfrak{a} \in I_n'$, let $o_n(\mathfrak{a})$ be the projection of $\underline{o}_n$ on

$$\mathrm{Hom}_k^c(A^2_{\underline{d}_o}(\underline{d}, O_{\underline{d}}), ((\ker h_{n+1}' + \mathfrak{a})/\mathfrak{a}))$$

and let $\mathfrak{a}'$ be the ideal of H_n' containing $\mathfrak{a}$ and such that as an ideal of $H_n'/\mathfrak{a}$, $\mathfrak{a}'/\mathfrak{a}$ is generated by the image of $o_n(\mathfrak{a})$ in $(\ker h_{n+1}'+\mathfrak{a})/\mathfrak{a}$. Obviously $\mathfrak{a} \subseteq \mathfrak{b}$ implies $\mathfrak{a}' \subseteq \mathfrak{b}'$.

Put

$$H_{n+1} = \lim_{\substack{\leftarrow \\ I_n'}} H_n'/\mathfrak{a}'$$

$$h_{n+1} : H_{n+1} \to H_n .$$

Let

$$\underline{e}_n^{n+1} \quad (\text{resp. } \underline{e}_{on}^{n+1})$$

denote the subcategory of 3.V-$\underline{\text{alg}}$. analoguous to the subcategory $\underline{e}_n'$ (resp. e_{on}') defined above, with H_m' and h_{m+1}' replaced by H_{n+1} and h_{n+1} respectively.

By construction the obstruction for deforming $\underline{e}_n^{n+1}$ relative to $\underline{e}_{on}^{n+1}$ vanish, therefore there exists a deformation σ_{n+1} of $\underline{e}_n^{n+1}$ relative to $\underline{e}_{on}^{n+1}$. Obviously σ_{n+1} is a deformation of $\underline{e}_{n+1}$ relative to $\underline{e}_{on+1}$. Moreover, by construction H_m and σ_m have the properties 1) and 2) for $m \leq n+1$.

Before we prove 3) for $m \leq n+1$, we shall study the ideals C_m and C_m' , and their relations to the family of homomorphism $o_m(\mathfrak{a})$, $\mathfrak{a} \in I_m$.

By construction we have:

$$C_n' = t_{n+1}^{-1}(C_n) \cdot \underline{m}_{T^1_{n+1}}$$

$$\ker h_{n+1}' = t_{n+}^{-1}(C_n)/C_n'$$

$$C_{n+1}/C_n' \subseteq t_{n+1}^{-1}(C_n)/C_n' .$$

These relations imply:

$$\underline{m}_{T^1_{n+1}} \cdot C_{n+1} \subseteq C_n'$$

which again implies that the morphism

$$t_{n+2} : T^1_{n+2} \to T^1_{n+1}$$

maps C_{n+1}' into C_n'.

There follows a commutative diagram

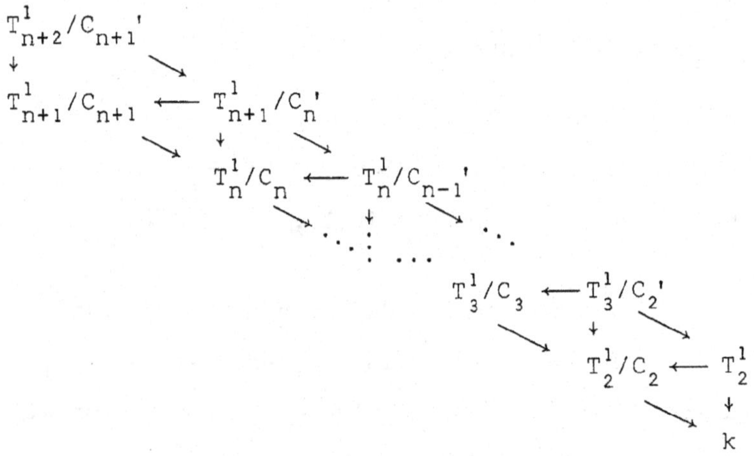

Denote by J_m the ordered set of open ideals of T^1_m. Using (4.2.3) on the parallellograms with two vertical edges, we find that the homomorphism

$$\varprojlim_{J_{n+1}} \mathrm{Hom}_k^c(A_{\underline{d}_0}(\underline{d},O_{\underline{d}})^*, (t^{-1}_{n+1}(C_n)+\mathfrak{a})/(C_n'+\mathfrak{a}))$$

$$\downarrow$$

$$\varprojlim_{J_n} \mathrm{Hom}_k^c(A^2_{\underline{d}_0}(\underline{d},O_{\underline{d}})^*, (t^{-1}_n(C_{n-1})+\mathfrak{a})/(C_{n-1}'+\mathfrak{a})$$

maps $\underline{o}_n$ onto $\underline{o}_{n-1}$.

Notice that we know already that C_2 is generated as ideal by the image of $\underline{o}_1$ considered as an element of

$$\varprojlim_{J_2} \mathrm{Hom}_k^c(A^2_{\underline{d}_o}(\underline{d},O_{\underline{d}})^*, \underline{m}_{T_2^1}/\mathfrak{a})$$
$$= \mathrm{Hom}_k^c(A^2_{\underline{d}_o}(\underline{d},O_{\underline{d}})^*, \underline{m}_{T_2^1}) .$$

Put

$$\underline{t}_{T^i} = \underline{m}_{T^i}/\underline{m}^2_{T^i} = A^i_{\underline{d}_o}(\underline{d},O_{\underline{d}})^*, \qquad i \geq 0$$

and consider the commutative diagram:

$$\begin{array}{ccc}
0 & & \\
\downarrow & & \\
\varprojlim_{J_{n+1}} \mathrm{Hom}_k^c(\underline{t}_{T^2},(C_n'+\mathfrak{a})/\mathfrak{a}) & \xrightarrow{t_{n+1}} & \varprojlim_{J_n} \mathrm{Hom}_k^c(\underline{t}_{T^2},(C_{n-1}'+\mathfrak{a})/\mathfrak{a}) \to 0 \\
\downarrow & & \downarrow \\
\varprojlim_{J_{n+1}} \mathrm{Hom}_k^c(\underline{t}_{T^2},(t_{n+1}^{-1}(C_n)+\mathfrak{a})/\mathfrak{a}) & \xrightarrow{k_{n+1}} & \varprojlim_{J_n} \mathrm{Hom}_k^c(\underline{t}_{T^2},(t_n^{-1}(C_{n-1})+\mathfrak{a})/\mathfrak{a}) \\
\downarrow r_{n+1} & & \downarrow r_n \\
\varprojlim_{J_{n+1}} \mathrm{Hom}_k^c(\underline{t}_{T^2},(t_{n+1}^{-1}(C_n)+\mathfrak{a})/(C_n'+\mathfrak{a})) & \to & \varprojlim_{J_n} \mathrm{Hom}_k^c(\underline{t}_{T^2},(t_n^{-1}(C_{n-1})+\mathfrak{a})/(C_{n-1}'+\mathfrak{a})) \\
\downarrow & & \downarrow \\
0 & & 0
\end{array}$$

All sequences in this diagram are exact. In fact, since J_{n+1} and J_n contain countable cofinal subsets, the only point to prove is that the upper horizontal sequence is exact. Now this is a consequence of a lemma of Mittag-Leffler type (see (La 2) (1.)) and the following equality:

$$t_{n+1}(C_n') = C_{n-1}' ,$$

which easily is seen to follow from the corresponding equality

$$t_n(C_n) = C_{n-1} .$$

We shall prove this by induction on n, knowing, of course, that it is true for $n = 2$.

Suppose there exist elements

$$O_m \in \varprojlim_{J_m} \mathrm{Hom}_k^c(\underline{t}_{T^2},(t_m^{-1}(C_{m-1})+\mathfrak{a})/\mathfrak{a})$$

$$\subseteq \mathrm{Hom}_k^c(\underline{t}_{T^2},T_m^1) \qquad m \le n$$

such that

$$k_m(O_m) = O_{m-1}, \qquad m \le n$$

$$r_n(O_m) = \underline{o}_{m-1}, \qquad m \le n$$

and such that C_m is generated by the image of O_m for $n \le n$, then obviously

$$t_m(C_m) = C_{m-1}, \qquad m \le n .$$

In particular, therefore, the upper horizontal sequence in the diagram above is exact.

But then, by elementary diagram chasing, we find an element

$$O_{n+1} \in \varprojlim_{J_{n+1}} \mathrm{Hom}_k^c(\underline{t}_{T^2},(t_{n+1}^{-1}(C_n)+\mathfrak{a})/\mathfrak{a}))$$

such that:

$$k_{n+1}(O_{n+1}) = O_n$$

$$r_{n+1}(O_{n+1}) = \underline{o}_n .$$

Knowing, as we do, that C_{n+1}/C_n' is generated by the image of $\underline{o}_n$ in $t_{n+1}^{-1}(C_n)/C_n'$, and recalling that

$$C_n' = t_{n+1}^{-1}(C_n)\cdot \underline{m}_{T^2_{n+1}}$$

we conclude that C_{n+1} is generated by the image of O_{n+1}.

Summing up, we have proved that there exists an element

$$o = \{O_n\}_{n\ge 1} \in \varprojlim_{n} \mathrm{Hom}_k^c(\underline{t}_{T^2},T_n^1) = \mathrm{Hom}_k^c(\underline{t}_{T^2},T^1) = \mathrm{Mor}^c(T^2,T^1)$$

such that

$$H = \varprojlim_{n} H_n = T^1 \hat{\otimes}_{T^2} k .$$

Moreover, in the process, we have proved that

$$H_{n+1}/\underline{m}^n_{H_{n+1}} = H_n , \qquad n \geq 1 .$$

To complete the proof of the theorem, we have to prove that (H_m, σ_m) has the property 3) .

Take any object R of $\underline{l}_{n+1}$ and consider the commutative diagram

$$\begin{array}{ccc}
\mathrm{Mor}^c(H_{n+1}, R) & \xrightarrow{\psi_{n+1}(R)} & \mathrm{Def}(\underline{d}/\underline{d}_o)(R) \\
\downarrow & & \downarrow \\
\mathrm{Mor}^c(H_n, R/\underline{m}^n_R) & \xrightarrow{\psi_n(R/\underline{m}^n_R)} & \mathrm{Def}(\underline{d}/\underline{d}_o)(R/\underline{m}^n_R)
\end{array}$$

which exists since we have proved that $H_{n+1}/\underline{m}^n_{H_{n+1}} = H_n$.

Now, use (4.2.1) to see that this diagram may be completed by the following commutative diagram:

$$\begin{array}{ccc}
\mathrm{Mor}^c(H_{n+1}, R) \times A^1_{\underline{d}_o}(\underline{d}, O_{\underline{d}} \otimes \underline{m}^n_R) & \to & \mathrm{Def}(\underline{d}/\underline{d}_o)(R) \times A^1_{\underline{d}_o}(\underline{d}, O_{\underline{d}} \otimes \underline{m}^n_R) \\
\| & & \| \\
\mathrm{Mor}^c(H_{n+1}, R) \times \mathrm{Mor}^c(H_{n+1}, k[\underline{m}^n_R]) & \to & \mathrm{Def}(\underline{d}/\underline{d}_o)(R) \times \mathrm{Def}(\underline{d}/\underline{d}_o)(k[\underline{m}^n_R]) \\
\downarrow \mu_1 & & \downarrow \mu_2 \\
\mathrm{Mor}^c(H_{n+1}, R) \times \mathrm{Mor}^c(H_{n+1}, R) & \to & \mathrm{Def}(\underline{d}/\underline{d}_o)(R) \times \mathrm{Def}(\underline{d}/\underline{d}_o)(R) \\
\mathrm{Mor}^c(H_{n+1}, R/\underline{m}^n_R) & & \mathrm{Def}(\underline{d}/\underline{d}_o)(R/\underline{m}^n_R)
\end{array}$$

Note that since H_2 prorepresents $\mathrm{Def}(\underline{d}/\underline{d}_o)$ on $\underline{l}_2$ the second projection of the upper horizontal morphism is an isomorphism.

We know by the induction hypotheses that $\psi_n(R/\underline{m}^n_R)$ is surjective.

Let us prove that this implies that $\psi_{n+1}(R)$ is surjective. Let, to that end, $\bar{\sigma}_R$ be any element of $\mathrm{Def}(\underline{d}/\underline{d}_o)(R)$ and let

$$\bar{\sigma}_{R/\underline{m}_R^n} = \bar{\sigma}$$

be the image of $\bar{\sigma}_R$ in $\mathrm{Def}(\underline{d}/\underline{d}_o)(R/\underline{m}_R^n)$. There exists a morphism of topological rings

$$\varphi : H_n \to R/\underline{m}_R^n$$

such that $\psi_n(R/\underline{m}_R^n)(\varphi) = \bar{\sigma}$. This of course means that

$$\bar{\sigma} = \mathrm{Def}(\underline{d}/\underline{d}_o)(\varphi)(\sigma_n) .$$

Consider the diagram

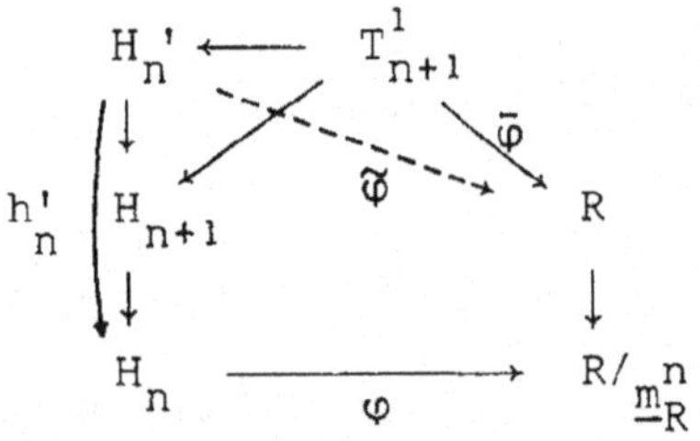

We may clearly find a morphism $\bar{\varphi}$ making the diagram of solid arrows commutative. Since $\bar{\varphi}$ maps $\ker\{T^1_{n+1} \to H_n\} \cdot \underline{m}_{T^1_{n+1}}$ onto zero, $\bar{\varphi}$ factors through H_n'. Finally, since $\bar{\sigma} = \bar{\sigma}_{R/\underline{m}_R^n}$ may be lifted to R the induced morphism in cohomology maps the obstruction

$$\underline{o}_n = o(\sigma_n, h_n')$$

to zero (see (4.2.3)). Therefore $\tilde{\varphi}$ factors through H_{n+1}, and we obtain a commutative diagram

$$\begin{array}{ccc} H_{n+1} & \xrightarrow{\varphi'} & R \\ \downarrow & & \downarrow \\ H_n & \xrightarrow{\varphi} & R/\underline{m}_R^n \end{array} .$$

This together with the nice diagram above, in which μ_1 and μ_2 are surjective, completes the proof of the theorem. QED.

Remark (4.2.5) By construction the obstruction morphism $o : T^2 \to T^1$ has the following properties

(i) $o(\underline{m}_{T^2}) \subseteq \underline{m}_{T^1}^2$

(ii) the leading term of o (the primary obstruction) is unique.

In particular (i) implies

$$A^1_{\underline{d}_o}(\underline{d}, O_{\underline{d}}) \simeq (\underline{m}_H/\underline{m}_H^2)^* .$$

Thus the imbedding dimension of H is equal to $\dim_k A^1_{\underline{d}_o}(\underline{d}, O_{\underline{d}})$.

Remark (4.2.6) The proof of (4.2.4) would be somewhat simpler if one assumes $\dim_k A^i < \infty$, $i = 1,2$, see the proof of Theorem (2.8) of [La 7].

(4.3) <u>The obstruction morphism and Massey products</u>

Consider the obstruction morphism

$$o : T^2 \to T^1$$

defined above.

We shall show that o is "essentially" determined by some cohomology operations defined on $A^{\cdot}_{\underline{d}_o}(\underline{d},0_{\underline{d}})$ which have the same properties as the classical Massey products of algebraic topology, see [May].

To avoid technical problems we shall restrict ourselves to the case where

$$V = k \quad \text{is of characteristic } 0$$

$$\dim_k A^i_{\underline{d}_o}(\underline{d},0_{\underline{d}}) < \infty \qquad i = 1,2 .$$

Put $A^i = A^i_{\underline{d}_o}(\underline{d},0_{\underline{d}})$, and consider the restriction M^* of o to $A^{2*} \simeq \underline{m}_{T^2}/\underline{m}_{T^2}^2 \subseteq T^2$.

Then M^* is a morphism of k-vector-spaces,

$$M^* : A^{2*} \to \prod_{n\geq 2} \overset{n}{\underset{\text{sym}}{\otimes}} A^{1*} .$$

Projecting onto the product of the first $r-1$ factors we obtain a morphism of k-vector-spaces

$$M^r : A^{2*} \to \prod_{n\geq 2}^{r} \overset{n}{\underset{\text{sym}}{\otimes}} A^{1*}$$

the dual of which is a morphism of k-vector-spaces, see (4.3.4)

$$M_r : \coprod_{n\geq 2}^{r} \overset{n}{\underset{\text{sym}}{\otimes}} A^1 \to A^2 .$$

Let $\{x_1,\dots,x_j,\dots,x_{d_1}\}$ be a basis of A^1 and denote by $\{x_1^*,\dots,x_j^*,\dots,x_{d_1}^*\}$ the dual basis of A^{1*}. In the same way, let $\{y_1,\dots,y_i,\dots,y_{d_2}\}$ be a basis of A^2 and denote by $\{y_1^*,\dots,y_i^*,\dots,y_{d_2}^*\}$ the dual basis of A^{2*}.

Obviously the composition o_{r+1} of o and the canonical morphism $T^1 \to T^1_r = T^1/\underline{m}_{T^1}^{r+1}$ is uniquely determined by M^r which in its turn is determined by the values

$$M^r(y_i^*) = \sum_{j_1\le j_2} \sigma(\underline{j})\alpha^i_{j_1,j_2} x_{j_1}^* x_{j_2}^* + \dots + \sum_{j_1\le\dots\le j_r} \sigma(\underline{j})\ \alpha^i_{j_1,\dots,j_r}\ x_{j_1}^* \dots x_{j_r}^*$$

for $i = 1,\dots,d_2$, where $\sigma(\underline{j}) = \frac{r!}{n_1!\dots n_s!}$ with $j_1 = j_{n_1} < j_{n_1+1} = \dots = j_{n_1+n_2} < \dots$ $\dots < j_{n_1+\dots+n_{s-1}} = \dots = j_{n_1+\dots+n_s} = j_r$.

y definition we find:

$$y_i^*(M_r(x_{j_1}\otimes\dots\otimes x_{j_n})) = \alpha^i_{j_1,\dots,j_n} \qquad \text{or}$$

$$M_r(x_{j_1}\otimes\dots\otimes x_{j_n}) = \sum_{i=1}^{d_2} \alpha^i_{j_1,\dots,j_n} y_i .$$

Now, going back to the construction of o it is clear that M^* is not, in general, unique.

However M^* is unique modulo the ideal of T^1 generated by the products $A^{1*}\cdot \operatorname{im} M^*$.

Therefore the obvious composition

$$M_0^r : A^{2*} \to (\prod^r \underset{\text{sym}}{\overset{n}{\otimes}} A^{1*})/(A^{1*}\cdot \operatorname{im} M^{r-1})$$

is uniquely defined. Moreover it is clear that o is essentially determined by the coherent sequence of homomorphisms $\{M_0^r\}_{r\ge 2}$.

Put

$$D_r = ((\prod_{n\ge 2}^{r} \underset{\text{sym}}{\overset{n}{\otimes}} A^{1*})/(A^{1*}\cdot \operatorname{im} M^{r-1}))^* ,$$

then in particular $D_2 = A^1 \underset{\text{sym}}{\otimes} A^1$.

Dualizing M_0^r we obtain uniquely defined cohomology operations

$$M_r^0 : D_r \to A^2$$

which characterize the morphism o.

Definition (4.3.1) M_r^o will be called the Massey products of $A^{\cdot}$ defined by the obstruction calculus. If $a_1 \otimes \ldots \otimes a_n \in D_r$ we shall denote by $\langle a_1, \ldots, a_n \rangle$ the value of M_r^o on $a_1 \otimes \ldots \otimes a_n$.

Obviously M_r is an extension of M_r^o from D_r to $\prod_{n\geq 2}^{r} \overset{n}{\underset{\text{sym}}{\otimes}} A^1$.
Following the tradition we shall also talk about Massey products when refering to M_r. Thus the Massey products M_r have a certain indeterminacy for $r \geq 3$.
In fact it is easy to see that $a_1 \otimes a_2 \otimes a_3 \in D_3$ if and only if $a_i \cup a_j = M_2^o(a_i \otimes a_j) = 0$ for all $i \neq j$, $i,j = 1,2,3$.

In the study of these Massey products we shall be inspired by a paper of Dwyer, see [Dw].
Consider the local k-algebra

$$U(r) = k[Z_1, \ldots, Z_r]/(Z_1^2, \ldots, Z_r^2)$$

Let ε_i denote the image of z_i in $U(r)$. Obviously $U(r)$ is a k-vector-space of finite dimension, therefore $U(r)$ is an object of the category $\underline{1}$ (see (4.2)).
The maximal ideal $\underline{m}$ of $U(r)$ is generated by $\varepsilon_1, \ldots, \varepsilon_r$ and has a k-vector-space basis given by the products

$$\varepsilon_{\underline{j}} = \varepsilon_{j_1} \cdot \varepsilon_{j_2} \cdots \varepsilon_{j_s}$$

where $1 \leq j_1 < j_2 \cdots < j_s \leq r$, $1 \leq s \leq r$.
In particular the ideal $\underline{m}^r$ is generated by $\varepsilon_1 \cdot \varepsilon_2 \cdots \varepsilon_r$ as an ideal as well as a k-vector-space.

A morphism of local k-algebras

$$T^1 \to U(r)$$

is determined by a k-linear map

$$A^{1*} \to \underline{m}$$

therefore by a collection of elements

$$a^1_{\underline{i}} = a^1_{i_1,\ldots,i_s} \in A^1 \qquad 1 \le i_1 < \cdots < i_s \le r,\ 1 \le s \le r.$$

Given such a collection $\{a_{\underline{i}}\}$ of elements of A^1, let

$$[a_{\underline{i}}] : T^1 \to U(r)$$

be the corresponding morphism and denote by $\langle\langle a_{\underline{i}}\rangle\rangle$ the composition $o\,[a_{\underline{i}}] : T^2 \to U(r)$. Then $\langle\langle a_{\underline{i}}\rangle\rangle$ corresponds to a collection

$$b_{\underline{j}} = b_{j_1,\ldots,j_s} \qquad 1 \le j_1 < \cdots < j_s \le r,\ 1 \le s \le r$$

of elements of A^2, with $b_{\underline{j}} = 0$ if $s = 1$.
It is easily checked that

$$b_{\underline{j}} = \sum_{\varepsilon_{\underline{i}_1}\cdots\,\varepsilon_{\underline{i}_t} = \varepsilon_{\underline{j}}} M_r(a_{\underline{i}_1} \otimes \cdots \otimes a_{\underline{i}_t})$$

$$= \sum_{\varepsilon_{\underline{i}_1}\cdots\,\varepsilon_{\underline{i}_t} = \varepsilon_{\underline{j}}} \langle a_{\underline{i}_1}, \ldots, a_{\underline{i}_t}\rangle$$

with the conventions above.

Now, consider the morphism

$$\rho : U(r) \to U(r)/\underline{m}^r$$

We know that there is a (one-to-possibly-many) correspondence between the elements of

$$\mathrm{Def}(\underline{d}/\underline{d}_0)(U(r)/\underline{m}^r)$$

and those morphisms

$$T^1 \to U(r)/\underline{m}^r$$

which composed with o are trivial.

Therefore any collection $\{a_{\underline{i}}\}$ of elements of A^1 such that the composition

$$T^2 \overset{o}{\to} T^1 \xrightarrow{[a_{\underline{i}}]} U(r) \overset{\rho}{\to} U(r)/\underline{m}^r$$

is trivial corresponds to an element

$$(a_{\underline{i}}) \in \mathrm{Def}(\underline{d}/\underline{d}_o)(U(r)/\underline{m}^r) .$$

Since $\ker \rho^2 = (\underline{m}^r)^2 = 0$ we may consider the obstruction $o(a_{\underline{i}})$ for lifting $(a_{\underline{i}})$ to $U(r)$. This obstruction sits in $A^2_{\underline{d}_o}(\underline{d}, O_{\underline{d}} \otimes_k \ker \rho) = A^2$ since $\ker \rho \simeq k$.

By construction of o, $o(a_{\underline{i}})$ corresponds to the restriction to A^{2*} of the composition $\langle\langle a_{\underline{i}} \rangle\rangle$. This restriction is, in fact a k-linear map

$$A^{2*} \to \underline{m}^r \simeq k .$$

Therefore, with the notations above

$$o(a_{\underline{i}}) = b_{1,2,\dots,r} = \sum_{\varepsilon_{\underline{i}_1} \cdots \varepsilon_{\underline{i}_t} = \varepsilon_{1,2,\dots,r}} \langle a_{\underline{i}_1}, \dots, a_{\underline{i}_t} \rangle$$

Summing up we find the following:

<u>Proposition (4.3.2)</u> Let $\{a_{\underline{i}}\}$ be a collection of elements of A^1 such that

$$\sum_{\varepsilon_{\underline{i}_1} \cdots \varepsilon_{\underline{i}_t} = \varepsilon_{\underline{j}}} \langle a_{\underline{i}_1}, \dots, a_{\underline{i}_t} \rangle = 0 \quad \text{for all } \underline{j} \neq (1,2,\dots,r) .$$

Then there is an element $(a_{\underline{i}}) \in \mathrm{Def}(\underline{d}/\underline{d}\ )(U(r)/\underline{m}^r)$ such that

$$o(a_{\underline{i}}) = \sum_{\varepsilon_{\underline{i}_1} \cdots \varepsilon_{\underline{i}_t} = \varepsilon_{1,2,\dots,r}} \langle a_{\underline{i}_1}, \dots, a_{\underline{i}_t} \rangle$$

This result is closely related to Theorem 2.4. of Dwyer, see [Dw].

Remark (4.3.3) Suppose $a_1 = \cdots = a_r$ and suppose further that $<a_1,\ldots,a_r>$ is defined. Then by (4.3.2)

$$<a_1,\ldots,a_r> = \frac{1}{r!}\, o(a_{\underline{i}})$$

where $(a_{\underline{i}})$ is the element of $\mathrm{Def}(\underline{d}/\underline{d}_o)(U(r)/\underline{m}^r)$ corresponding to the collection $\{a_i\}_{i=1}^{r}$.

In this case we may also proceed as follows. Put $S^n = k[t]/(t^n)$ and consider the k-linear map $a: A^{1*} \to (t)/(t^2) \subseteq S^2$ defined by the identification of $(t)/(t^2) \simeq k$. The induced morphism $T^1 \to S^2$ corresponds uniquely to an element $(a) \in \mathrm{Def}(\underline{d}/\underline{d}_o)(S^2)$. The obstruction $o(a) \in A^2_{\underline{d}_o}(\underline{d}; O_{\underline{d}} \otimes_k (t^2)/(t^3)) = a^2_{\underline{d}_o}(\underline{d}; O_{\underline{d}})$ for lifting (a) to S^3 is equal to $<a,a>$. Thus if $<a,a> = 0$ there is an element $(a)_3 \in \mathrm{Def}(\underline{d}/\underline{d}_o)(S^3)$ lifting (a). Then $<a,a,a>$ is the obstruction for lifting $(a)_3$ to S^4 etc.

The morphism $S^{r+1} \to U(r)$ defined by sending t to $\sum_{i=1}^{r} \varepsilon_i$ induces a map $\mathrm{Def}(\underline{d}/\underline{d}_o)(S^{r+1}) \to \mathrm{Def}(\underline{d}/\underline{d}_o)(U(r))$, which maps (a) to $(a_{\underline{i}})$ where $a_i = a$, $i = 1,\ldots,r$. However, the induced map $(t^r)/(t^{r+1}) \simeq k \to k \simeq \underline{m}^r$ is the multiplication by $r!$. This explains the formula above.

Remark (4.3.4) For any field k and any k-vector space V the symmetric groups S_n acts on $\overset{n}{\otimes} V$ by permuting factors. There is a well known commutative diagram

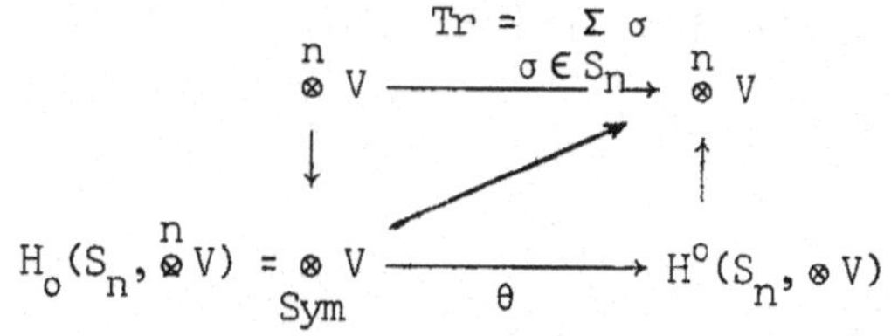

If the characteristic p of k is prime to $n!$ θ is an isomorphism and one usually identifies $\underset{\mathrm{Sym}}{\overset{n}{\otimes}} V$ and $H^o(S_n,\overset{n}{\otimes} V)$ via the isomorphism $\frac{1}{n!}\cdot\theta$.

Dualizing we find the diagram

$$\begin{array}{ccc} (\overset{n}{\otimes} V)^* = \overset{n}{\otimes} V^* & \xleftarrow{\;Tr = \sum_{\sigma\in S_n}\sigma\;} & \overset{n}{\otimes} V^* = (\overset{n}{\otimes} V)^* \\ \uparrow & & \downarrow \\ (\underset{\mathrm{Sym}}{\overset{n}{\otimes}} V)^* = H^o(S_n,\overset{n}{\otimes} V^*) & \xleftarrow[\theta^*]{} & \underset{\mathrm{Sym}}{\overset{n}{\otimes}} V^* = H_o(S_n,\overset{n}{\otimes} V^*) \end{array}$$

Therefore if $p \nmid n!$ there is a natural isomorphism

$$\frac{1}{n!}\,\theta^* : (\underset{\mathrm{Sym}}{\overset{n}{\otimes}} V)^* \xrightarrow{\sim} \underset{\mathrm{Sym}}{\overset{n}{\otimes}} V^* .$$

The computations above depend on these identifications.

Notice that without restrictions on the characteristic the Massey products M_r^o are defined on subspaces D_r of $\coprod_{n=2}^{r} H^o(S_n,\overset{n}{\otimes} A^1)$ defined inductively by:

$$D_2 = H^o(S_2,\overset{2}{\otimes} A^1) .$$

Given $a^* \in A^{1*}$ consider the contraction

$$a^* : H^o(S_n,\overset{n}{\otimes} A^1) \to H^o(S_{n-1},\overset{n-1}{\otimes} A^1)$$

defined by letting a^* operate on the first factor. Then

$$D_r = \{\alpha \in \coprod_{n=2}^{r} H^o(S_n,\overset{n}{\otimes} A^1) \mid \forall a^* \in A^{1*}, a^*\alpha \in D_{r-1} \text{ and } M_{r-1}^o(a^*\alpha) = 0\}.$$

This follows immediately from the definitions above.

Chapter 5

Some applications

In this chapter we shall study deformations of a morphism of algebraic k-schemes $f : X \to Y$. When f is affine we compute the first 3 cohomology groups and the Massey products. Using the results of (3.2) and the computations in the affine case we are able to compute the cohomology and the Massey products for any nice projective variety X, see (5.2).

When f is a closed imbedding of projective k-schemes the same procedure works, properly relativized. Thus we find a method for computing the completion of the local ring of $Hilb_Y$ at the point corresponding to $f : X \to Y$, provided X and Y are sufficiently well behaved, see (5.2).

We end this chapter with a study of the case where $X = Spec(k)$, $Y = Spec(A)$ and A is a local k-algebra. The resulting theorem has some interesting analogies in other fields of mathematics.

For further applications of the general theory see [La6] and [Lø].

(5.1) Local structure of some moduli schemes

Let k be any field and consider a morphism of algebraic k-schemes

$$f : X \to Y .$$

Denote by $k/sch/k$ the category of pointed k-schemes and let the functor

$$D_f : k/\underline{sch}/k \to \underline{Sets}$$

be defined by

$$D_f(\mathrm{Spec}(k) \underset{\varphi}{\to} T) = \left\{ \begin{array}{ccc} X' & \xrightarrow{f'} & Y \times T \\ \uparrow & \square & \uparrow 1\times\varphi \\ X & \xrightarrow[f]{} & Y \end{array} \;\middle|\; X' \text{ flat over } T \right\} \Big/ \sim$$

Here $\sim$ denotes the equivalence relation defined by

$$\begin{array}{ccc} X' & \xrightarrow{f'} & Y \times T \\ \uparrow & \square & \uparrow 1\times\varphi \\ X & \xrightarrow[f]{} & Y \end{array} \quad \sim \quad \begin{array}{ccc} X'' & \xrightarrow{f''} & Y \times T \\ \uparrow & \square & \uparrow 1\times\varphi \\ X & \xrightarrow[f]{} & Y \end{array}$$

if and only if there is an isomorphism $X' \to X''$ making all diagrams commute. The little square is shorthand for cartesian diagram.

In general there is no reason why D_f should be representable. However, the restriction of D_f to the subcategory $\underline{\ell}^o$ of $k/\underline{sch}/k$ is the deformation functor $\mathrm{Def}(\underline{d}_f)$, where $\underline{d}_f$ is the category of morphisms of k-algebras induced by f, thus D_f has a hull provided $\dim_k A^i(k, \underline{d}_f; O_{\underline{d}_f}) = \dim_k A^i(k,f;O_X)$ is countable for $i = 1.2$. Since X and Y are of finite type over k these conditions hold and we find

<u>Theorem (5.1.1)</u> Under the conditions above D_f has a hull H given by

$$H \simeq \mathrm{Sym}_k(A^1(f;O_X)^*)^\wedge \underset{\mathrm{Sym}_k(A^2(f;O_X)^*)^\wedge}{\hat{\otimes}} k \quad .$$

where $\mathrm{Sym}_k(A^1(f;O_X)^*)^\wedge$ is a $\mathrm{Sym}_k(A^2(f;O_X)^*)^\wedge$ - module via the obstruction morphism

$$o : \mathrm{Sym}_k(A^2(f;O_X)^*)^\wedge \to \mathrm{Sym}_k(A^1(f;O_X)^*)^\wedge .$$

In particular the imbedding dimension of H is $\dim_k A^1(f;O_X)$ and H is non-singular if and only if o is trivial.

Proof. This follows immediately from (4.1.9) and (4.2.4) together with the remark (4.2.5). QED.

Remark (5.1.2) Suppose D_f is representable, and let

$$m : \mathrm{Spec}(k) \to M$$

be the representing object of $k/\underline{sch}/k$. Then H is the hull of the functor $\mathrm{Mor}(-,M)$ restricted to $\underline{\ell}^o$. But $\mathrm{Mor}(-,M)$ restricted to $\underline{\ell}^o$ is dual to the functor $\mathrm{Mor}(\hat{O}_{M,m},-)$ defined on $\underline{\ell}$. Therefore

$$H \simeq \hat{O}_{M,m} .$$

Suppose now $f : X = \mathrm{Spec}(A) \to Y = \mathrm{Spec}(S)$ is a morphism of affine schemes. Then

$$A^i(f,O_X) = A^i(S,A;A) \qquad i \geq 0 ,$$

and the hull of the functor D_f is given by

$$H \simeq \mathrm{Sym}_k(H^1(S,A;A)^*)^\wedge \hat{\otimes}_{\mathrm{Sym}_k(H^2(S,A;A)^*)^\wedge} k$$

Our first task is to compute $H^i(S,A;A)$, for $i = 1,2$, and the Massey products.

Consider the resolution of (2.1),

$$A \underset{\rho}{\leftarrow} F_0 \leftleftarrows F_1 \ \overleftarrow{\overleftarrow{\leftarrow}} \cdots \ \overleftarrow{\vdots} \ F_p \ \overleftarrow{\vdots} \ \cdots$$

where F_0 is S-free , ρ is epi and

$$F_p = \underbrace{F_0 \times_A \ldots \times_A F}_{p+1}$$

We have seen that the cohomology of the double complex

$$C^{\cdot}(S\text{-}\underline{\text{free}}/F_{\cdot}^{o}, \mathrm{Der}_k(-,M))$$

is equal to $H^{\cdot}(S,A;M)$, for any A-module M .

The Leray spectral sequence, see (2.1), is given by

$$E_2^{p,q} = H^p(H^q(S,F_{\cdot};M)) .$$

Identifying F_p as S-module with $F_0 \oplus \underbrace{J \oplus \ldots \oplus J}_{p}$ where $J = \ker \rho$, it is easy to see that an S-derivation $D \in \mathrm{Der}_S(F_p,M)$ corresponds uniquely to a p+1 - tuple $(D_o,h_1,\ldots,h_p)$ where $D_o \in \mathrm{Der}_S(F_0,M)$, $h_i \in \mathrm{Hom}_{F_0}(J,M)$

Using this we find, after a dull computation

$$\begin{aligned}
E_2^{o,o} &= \mathrm{Der}_S(A,M) \\
E_2^{o,1} &= 0 \\
H^1(S,A;M) = E_2^{1,o} &= \mathrm{Hom}_{F_0}(J,M)/_{\mathrm{Der}} \\
E_2^{2,o} &= 0 \\
E_2^{3,o} &= 0 \\
E_2^{o,2} &= 0 \\
H^2(S,A;M) = E_2^{1,1} &= \ker(H^1(S,F_0 \underset{A}{\times} F_0;M) \ \overleftarrow{\overleftarrow{\leftarrow}} \ H^1(S,F_0 \underset{A}{\times} F_0 \underset{A}{\times} F_0;M))
\end{aligned}$$

Let J be generated by $f_1,\dots,f_m$, and let

$$\varphi : F_o[x_1,\dots,x_m] \to F_o \underset{A}{\times} F_o = F_o \oplus J$$

be defined by $\varphi(x_i) = (0,f_i)$. Then φ is surjective. Putting $R = F_o[x_1,\dots,x_m]$ and $I = \ker\varphi$ we obtain

$$H^1(S,F_o \underset{A}{\times} F_o;M) = \mathrm{Hom}_R(I,M)/\mathrm{Der}$$

In the same way, defining

$$\psi : F_o[x_1,\dots,x_m,y_1,\dots,y_m] \to F_o \underset{A}{\times} F_o \underset{A}{\times} F_o = F_o \oplus J \oplus J$$

by $\psi(x_i) = (0,f_i,0)$, $\psi(y_j) = (0,0,f_j)$ we find

$$H^1(S,F_o \underset{A}{\times} F_o \underset{A}{\times} F_o;M) = \mathrm{Hom}_T(H,M)/\mathrm{Der}$$

where we have put

$$T = F_o[x_1,\dots,x_m,y_1,\dots,y_m]$$
$$H = \ker\psi .$$

Let $p_i : F_o \underset{A}{\times} F_o \underset{A}{\times} F_o \to F_o \underset{A}{\times} F_o, i=1,2,3$ be the projections gotten by disregarding the ith coordinate. Define morphisms

$$p_i' : F_o[x_1,\dots,x_m,y_1,\dots,y_m] \to F_o[x_1,\dots,x_m] \qquad i = 1,2,3$$

by:

$$\begin{array}{ll} p_1'(x_k) = f_k - x_k & p_1'(y_l) = x_l \\ p_2'(x_k) = 0 & p_2'(y_l) = x_l \\ p_3'(x_k) = x_k & p_3'(y_l) = 0 \end{array}$$

Then one easily checks that

$$\psi p_i = p_i' \varphi \qquad i = 1,2,3 .$$

Notice, in particular, that $x_k \cdot y_l \in H$ for all k,l and that

$$\sum_{i=1}^{3} (-1)^i p_i'(x_k \cdot y_l) = (f_k - x_k)x_l \in I .$$

Since any derivation $D \in \mathrm{Der}_S(T,M)$ maps $x_k \cdot y_l$ onto zero for $k,l = 1,\dots,m$, we find, summing up

<u>Proposition (5.1.3)</u> Let $I_1 \subseteq R$ be the F_0-module of linear relations among the $f_1,\dots,f_m$'s and let $I_o \subseteq I_1$ be the sub F_0-module generated by the trivial relations $f_k x_l - f_l x_k$, $k,l = 1,\dots,m$. Then

$$H^2(S,A;M) = \mathrm{Hom}_{F_0}(I_1/I_o,M)/\mathrm{Der}\,.$$

<u>Corollary (5.1.4)</u> Let A be a local k-algebra with maximal ideal $\underline{m}$. Suppose $k = A/\underline{m}$, then $H^2(k,A;k) = 0$ if and only if A is a complete intersection.

<u>Proof</u>. One checks that in this case $H^2(k,A;k) = \mathrm{Hom}_{F_0}(I_1/I_o,k)$.

QED.

To compute the Massey products we shall have to identify elements of $H^i(S,A;M)$, given in terms of cocycles of $C^{\cdot}(S\text{-}\underline{\text{free}}/A^o,\mathrm{Der}_S(-,M))$, with equivalence classes of homomorphisms in $\mathrm{Hom}_R(J,M)$ and $\mathrm{Hom}_{F_0}(I_1/I_o,M)$ for $i=1$ and $i=2$ respectively.

Let $\xi = \{\xi_{\psi_1}\} \in C^1(S\text{-}\underline{\text{free}}/A^o,\mathrm{Der}_S(-,M))$ be a 1-cocycle and consider the diagram

$$\begin{array}{lll}
C^1(S\text{-}\underline{\text{free}}/A^o,\mathrm{Der}_S(-,M)) & & \\
\rho_* \downarrow & & \\
C^1(S\text{-}\underline{\text{free}}/F_0^{\,o},\mathrm{Der}_S(-,M)) & \xleftarrow{d_o^o} & C^o(S\text{-}\underline{\text{free}}/F_0^{\,o},\mathrm{Der}_S(-,M)) \\
& & p_{1*}\downarrow\downarrow p_{2*} \\
& \xleftarrow{d_1^o} & C^o(S\text{-}\underline{\text{free}}/F_1^{\,o},\mathrm{Der}_S(-,M))
\end{array}$$

Since F_0 is S-free there exists a 0-cochain $\eta = \{\eta_\delta\}$ of $C^o(S\text{-}\underline{free}/F_0^o, Der_S(-,M))$ such that $d_0^o(\eta) = \rho_*(\xi)$. In fact, for every object $\delta : F' \to F_0$ of $S\text{-}\underline{free}/F_0$ we may define η by $\eta_\delta = \xi(\delta)$. The 0-cochain $p_{1*}(\eta) - p_{2*}(\eta)$ is therefore a 0-cocycle of $C^\cdot(S\text{-}\underline{free}/F_1^o, Der_S(-,M))$ and defines an element of $Der_S(F_1,M)$ which for obvious reasons is mapped to zero in $Der_S(F_2,M)$ by the map $p_{1*} - p_{2*} + p_{3*}$. Therefore it defines an element of $Hom_{F_0}(J,M)$ which in turn represents an element of $Hom_{F_0}(J,M)/Der$.

The morphism $\varphi : F_0[x_1,\dots,x_m] \to F_1$ above, is an object of $S\text{-}\underline{free}/F_1$, and $\varphi p_i : F_0[x_1,\dots,x_m] \to F \quad i = 1,2$ are objects of $S\text{-}\underline{free}/F_0$. Moreover $(p_{1*}-p_{2*})(\eta)_\varphi = \eta_{\varphi p_1} - \eta_{\varphi p_2} = \xi(\varphi p_1) - \xi(\varphi p_2)$.

Therefore the cocycle ξ corresponds to the element of $Hom_{F_0}(J,M)/Der$ represented by the homomorphism $h \in Hom_{F_0}(J,M)$ defined by

$$h(f_i) = (\xi(\varphi p_1) - \xi(\varphi p_2))(x_i).$$

In the same way, consider the diagram

$$\begin{array}{lcl}
C^2(S\text{-}\underline{free}/A^o;Der_S(-,M)) & & \\
\rho_* \downarrow & & \\
C^2(S\text{-}\underline{free}/F_0^o;Der_S(-,M)) & \xleftarrow{d_0^1} & C^1(S\text{-}\underline{free}/F_0^o;Der_S(-,M)) \\
 & & p_{1*}\uparrow\uparrow p_{2*} \\
 & & C^1(S\text{-}\underline{free}/F_1^o;Der_S(-,M)) .
\end{array}$$

Let $\lambda = \{\lambda(\psi_1,\psi_2)\} \in C^2(S\text{-}\underline{free}/A^o;Der_S(-,M))$ be a cocycle representing an element of $H^2(S,A;M)$. As above it is easy to see that there is a 1-cochain $\nu = \{\nu(\psi_1)\} \in C^1(S\text{-}\underline{free}/F_0^o;Der_S(-,M))$ such that $\rho_*(\lambda) = d_0^1(\nu)$. In fact ν may be defined in the following way.

Let

$$\begin{array}{ccc} F' & \xrightarrow{\psi_1} & F'' \\ & \delta_1 \searrow \quad \swarrow \delta_2 & \\ & F_o & \end{array}$$

be any morphism of $S\text{-}\underline{\text{free}}/F_0$, put $\nu(\psi_1) = \lambda(\psi_1,\delta_2)$.

The 1-cocycle $p_{1*}(\nu) - p_{2*}(\nu)$ of $C^1(s\text{-}\underline{\text{free}}/F_1^o;\mathrm{Der}_S(-,M))$ therefore defines an element of $H^1(S,F_1;M)$ which is mapped to zero in $H^1(S,F_2;M)$ by the map $p_{1*} - p_{2*} + p_{3*}$. Therefore $p_{1*}(\nu) - p_{2*}(\nu)$ is an element of $H^2(S,A;M) = \mathrm{Hom}_{F_0}(I_1/I_o,M)/\mathrm{Der}$

Let $r_1,\ldots,r_n$ be generators of the ideal $I = \ker\varphi$, and let

$$\phi : F_0[x_1,\ldots,x_m,z_1,\ldots,z_n] \to F_0[x_1,\ldots,x_m] \underset{F_0}{\times} F_0[x_1,\ldots,x_m]$$

be defined by $\phi(x_i) = (x_i,x_i)$, $\phi(z_j) = (0,r_j)$.

Then the 1-cocycle $p_{1*}(\nu) - p_{2*}(\nu)$ defines a homomorphism $g \in \mathrm{Hom}_T(I,M)$ determined by

$$g(r_j) = [(p_{1*}(\nu) - p_{2*}(\nu))(\phi p_1') - (p_{1*}(\nu) - p_{2*}(\nu))(\phi p_2')](z_j)$$

where

$$p_i' : F_0[x_1,\ldots,x_m] \underset{F_1}{\times} F_0[x_1,\ldots,x_m] \to F_0[x_1,\ldots,x_m] \qquad i = 1,2$$

are the projections. This implies

$$g(r_j) = [\nu(\phi p_1',\varphi p_1) - \nu(\phi p_1',\varphi p_2) - \nu(\phi p_2',\varphi p_1) + \nu(\phi p_2',\varphi p_2)](z_j) .$$

With this done we may compute the cup-product

$$\cup : H^1(S,A;A) \underset{\mathrm{Sym}}{\otimes} H^1(S,A;A) \to H^2(S,A;A) ,$$

assuming for simplicity that the characteristic of the field k is different from 2 . It suffices to compute the cup-powers. Therefore, let $h \in \mathrm{Hom}_{F_0}(J,A)$ represent an element α of $H^1(S,A;A)$. We want to represent $\alpha \cup \alpha = \langle\alpha,\alpha\rangle \in H^2(S,A;A)$ as a homomorphism $g \in \mathrm{Hom}_T(I,M)$.

We know that α determines a lifting A' of A to $S_2 = S[t]/(t^2)$ given in the following way.
Let, with the notations above, $h_i' = h'(f_i) \in F_0$ be arbitrarily chosen elements such that

$$\rho(h_i') = h(f_i) .$$

Then A' is the quotient of $F_0^{(2)} = S_2 \underset{S}{\otimes} F_0$ by the ideal J' generated by $f_1^{(2)},\dots,f_m^{(2)}$ where

$$f_i^{(2)} = f_i + t\cdot h_i' .$$

Consider the morphisms

$$\varphi^{(2)} : F_0^{(2)}[x_1,\dots,x_m] \to F_0^{(2)} \underset{A'}{\times} F_0^{(2)} = F_1^{(2)}$$

defined by $\varphi^{(2)}(x_i) = (0,f_i^{(2)})$ $i=1,\dots,m$, and

$$\phi^{(2)} : F_0^{(2)}[x_1,\dots,x_m, z_1,\dots,z_n] \to F_0^{(2)}[x_1,\dots,x_m] \underset{F_1^{(2)}}{\times} F_0^{(2)}[x_1,\dots,x_m]$$

defined by $\phi^{(2)}(x_i) = (x_i,x_i)$, $\phi^{(2)}(z_j) = (0,r_j)$ where the $r_1^{(2)},\dots,r_n^{(2)}$ generate $I' = \ker\varphi^{(2)}$. It is easy to see that $r_j^{(2)} = r_j + t\, r_j'$ where

$$r_j'(f_1,\dots,f_m) = -r(h_1',\dots,h_m') .$$

Remember that r_j and r_j' are polynomials in $x_1,\dots,x_m$.
As we know, the cup-product $<\alpha,\alpha>$ is the obstruction $o(A') \in H^2(S_2,A';A' \underset{S_2}{\otimes} (t^2)) = H^2(S,A;A)$ for lifting A' to $S_3 = S[t]/(t^3)$. This obstruction is given by a 2-cocycle ν of $C^{\cdot}(S_2\text{-}\underline{\text{free}}/A'^{o}, \mathrm{Der}_{S_2}(-,A' \underset{S_2}{\otimes} (t^2)))$ defined as follows. Let

$$\begin{array}{ccccc} F_0'' & \xrightarrow{\psi_1} & F_1'' & \xrightarrow{\psi_2} & F_2'' \\ & {\scriptstyle \delta_o''}\searrow & \downarrow{\scriptstyle \delta_1''} & \swarrow{\scriptstyle \delta_2''} & \\ & & A' & & \end{array}$$

be a composition of morphisms of S_2-free/A' then given any quasisection $\sigma'' : S_2\text{-}\underline{\text{free}}/A' \to S_3\text{-}\underline{\text{free}}$ we have

$$\nu(\psi_1,\psi_2) = (\sigma''(\psi_1 \circ \psi_2) - \sigma''(\psi_1)\sigma''(\psi_2))\delta_2'' .$$

Consider in particular the diagram of morphisms of S_2-free/A'

$$\begin{array}{ccccc} F_o^{(2)}[x_1,\ldots,x_m,z_1,\ldots,z_n] & \underset{\phi^{(2)}p_2'}{\overset{\phi^{(2)}p_1'}{\rightrightarrows}} & F_o^{(2)}[x_1,\ldots,x_m] & \underset{\varphi^{(2)}p_2}{\overset{\varphi^{(2)}p_1}{\rightrightarrows}} & F_o^{(2)} \\ & \searrow & \downarrow & \swarrow & \\ & & A' & & \end{array}$$

It is easy to compute the 4 values

$$\nu(\phi^{(2)}p_i',\varphi^{(2)}p_j)(z_k) \qquad i,j = 1,2 .$$

In fact, let σ'' be the natural quasisection, defined by lifting the values of the indeterminants in the obvious way, then

$$\begin{aligned} \nu(\phi^{(2)}p_1',\varphi^{(2)}p_1)(z_k) &= -t^2 r_k'(h_1,\ldots,h_m) \\ &= t^2 h(r_k(h_1',\ldots,h_m')) . \\ \nu(\phi^{(2)}p_i',\varphi^{(2)}p_j)(z_k) &= 0 \quad \text{for } i \neq 1 \text{ or } j \neq 1 . \end{aligned}$$

Thus

$$g(r_k) = h(r_k(h_1',\ldots,h_m')) .$$

Summing up we have proved the following

<u>Proposition (5.1.5)</u> Let $\alpha \in H^1(S,A;A)$ be represented by the homomorphism $h \in \mathrm{Hom}_F(J,A)$, then the cup product $\langle\alpha,\alpha\rangle \in H^2(S,A;A)$ is represented by the homomorphism $g \in \mathrm{Hom}_T(I,A)$ where for any $r \in I$

$$g(r) = h(r(h_1',\ldots,h_m')) .$$

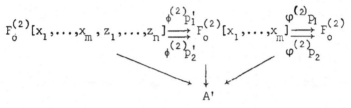

Suppose $\langle\alpha,\alpha\rangle = 0$. This means that $g \in \mathrm{Hom}_{F_o}(I_1/I_o, A)$ is a derivation, i.e. that there exist elements $k_i \in A$, $i = 1,\ldots,m$ such that

$$g(\sum_{i=1}^{m} r_j^i x_i) = \sum_{i=1}^{m} r_j^i k_i = r_j(k_1,\ldots,k_m)$$

whenever $r_j = \sum_{i=1}^{m} r_j^i x_i$ is a relation of I_1.

Let $k_i' \in F_o$, $i = 1,\ldots,m$ be arbitrary elements such that $\rho(k_i') = k_i$, $i = 1,\ldots,m$.

Put $S_n = S[t]/(t^n)$, $F_o^{(n)} = S_n \otimes_S F_o$, $n \geq 2$ and let

$$f_i^{(3)} = f_i + t\cdot h_i' + t^2 k_i', \qquad i = 1,\ldots,m.$$

Then $J^{(3)} \subseteq F_o^{(3)}$ generated by the $f_i^{(3)}$, $i = 1,\ldots,m$ defines a lifting of A' to S_3. Corresponding to a relation $r_j \in I_1$ there is a relation

$$r_j^{(3)} = r_j + t\, r_j' + t^2 r_j''$$

of the $f_i^{(3)}$'s $i = 1,\ldots,m$. In fact

$$r_j^{(3)}(f_1^{(3)},\ldots,f_m^{(3)}) = 0 \quad (\mathrm{mod}\ t^3)$$

is equivalent to

$$r_j'(f_1,\ldots,f_m) + r_j(h_1',\ldots,h_m') = 0$$

$$r_j''(f_1,\ldots,f_m) + r_j'(h_1',\ldots,h_m') + r_j(k_1',\ldots,k_m') = 0$$

therefore to

$$r_j'(f_1,\ldots,f_m) = -r_j(h_1',\ldots,h_m')$$

$$r_j''(f_1,\ldots,f_m) = -r_j'(h_1',\ldots,h_m') - r_j(k_1',\ldots,k_m').$$

We know already that such r_j' $j = 1,\ldots,n$ exist, and since $h(r_j(h_1',\ldots,h_m')) = r_j'(h_1,\ldots,h_m) = r_j(k_1,\ldots,k_m)$ for all $j = 1,\ldots,n$,

$-r_j'(h_1',\dots,h_m') - r_j(k_1',\dots,k_m') \in J$, therefore the r_j'', $j=1,\dots,n$ exist.

Just as above the Massey-product $\langle\alpha,\alpha,\alpha\rangle$ is now the obstruction of lifting

$$A'' = F_0^{(3)}/J^{(3)}$$

to S_4. In exactly the same way as above we prove that $\langle\alpha,\alpha,\alpha\rangle$ is represented by the homomorphism

$$g \in \mathrm{Hom}_{F_0}(I_1/I_0, A)$$

where

$$\begin{aligned} g(r_j) &= -r_j''(h_1,\dots,h_m) - r_j'(k_1,\dots,k_m) \\ &= h(r_j'(h_1',\dots,h_m') + r_j(k_1',\dots,k_m')) - r_j'(k\ ,\dots,k_m) \end{aligned}$$

It is clear that we may, in this way, compute any Massey product $\langle\alpha_1,\dots,\alpha_n\rangle$, for which there exists a defining system. Obviously we must in the general case replace the S_n by the corresponding $S \otimes_k U(n)$, but this leads only to minor changes in the computations.

<u>Example</u> (5.1.6) Consider the following example given by Pinkham in [Pi]. Let A be the cone of $\mathbb{P}^1$ imbedded in $\mathbb{P}^4$ by $\mathcal{O}_{\mathbb{P}^1}(4)$. Then $A = k[x_0,x_1,x_2,x_3,x_4]/(f_{02},f_{03},f_{04},f_{13},f_{14},f_{24})$ where $f_{ij} = x_i x_j - x_{i+1}x_{j-1}$ are the subdeterminants of the matrix

$$\begin{pmatrix} x_0 & x_1 & x_2 & x_3 \\ x_1 & x_2 & x_3 & x_4 \end{pmatrix}$$

There are 8 relations

$$
\begin{aligned}
R^0_{13} &= x_2f_{02} - x_1f_{03} + x_0f_{13}\\
R^0_{14} &= x_3f_{02} - x_1f_{04} + x_0f_{14}\\
R^0_{24} &= x_3f_{03} - x_2f_{04} + x_0f_{24}\\
R^1_{24} &= x_3f_{13} - x_2f_{14} + x_1f_{24}\\
S^0_{23} &= x_3f_{02} - x_2f_{03} + x_1f_{13}\\
S^0_{24} &= x_4f_{02} - x_2f_{04} + x_1f_{14}\\
S^0_{34} &= x_4f_{03} - x_3f_{04} + x_1f_{24}\\
S^1_{34} &= x_4f_{13} - x_3f_{14} + x_2f_{24}
\end{aligned}
$$

One computes and find

$$\dim_k A^1(k,A;A) = 4 .$$

A basis for $A^1(k,A;A)$ is represented by the following 4 homomorphisms A_0, A_1, A_2 and C_0 $\in \mathrm{Hom}_R((f_{02},f_{03},f_{04},f_{13},f_{14},f_{24}),A)$, given by the following matrix in which the element with coordinates (h,f), where h is one of the A_0,A_1,A_2,C_0 and f is one of the generators f_{ij}, is an element of $R = k[x_0,x_1,x_2,x_3,x_4]$ representing $h(f)$.

	f_{02}	f_{03}	f_{04}	f_{13}	f_{14}	f_{24}
A_0	$-x_1$	$-x_2$	$-x_3$	0	0	0
A_1	x_0	0	0	$-x_2$	$-x_3$	0
A_2	0	x_0	0	x_1	0	$-x_3$
C_0	$-x_0$	$-x_1$	$-x_2$	0	0	0

Let us compute the cup-product $C_0 \cup C_0$. This of course means that we want to compute $C_0 \cup C_0$ as a homomorphism $\mathrm{Hom}_R((R^k_{ij},S^k_{ij}),A)$.

Consider the, hopefully, self explanatory diagram

$C_o \cup C_o$		$R(h_1',\dots,h_6') = R'(f_{02},\dots,f_{24})$			$R'(h_1',\dots,h_6') = C_o \cup C_o(R)$
R_{13}^o	$\to$	$-x_2x_o + x_1^2$	$= f_{o2}$	$\to$	x_o
R_{14}^o	$\to$	$-x_3x_o + x_1x_2$	$= -f_{o3}$	$\to$	x_1
R_{24}^o	$\to$	$-x_3x_1 + x_2^2$	$= -f_{13}$	$\to$	0
R_{24}^1	$\to$	0	$= 0$	$\to$	0
S_{23}^o	$\to$	$-x_3x_o + x_2x_1$	$= -f_{o3}$	$\to$	x_1
S_{24}^o	$\to$	$-x_4x_o + x_2^2$	$= -f_{o4} - f_{13}$	$\to$	x_2
S_{34}^o	$\to$	$-x_4x_1 + x_3x_2$	$= -f_{14}$	$\to$	0
S_{34}^1	$\to$	0	$= 0$	$\to$	0

To see that $C_o \cup C_o$ as an element of $\mathrm{Hom}_R((R_{ij}^k, S_{ij}^k), A)$ is not an element of Der we first check the first values $C_o \cup C_o(R_{ij})$. Suppose $C_o \cup C_o \in$ Der, then we must have:

$$x_o = x_2k_1 - x_1k_2 + x_ok_4$$
$$x_1 = x_3k_1 - x_1k_3 + x_ok_5$$
$$0 = x_3k_2 - x_2k_3 + x_ok_6$$

for some $k_i \in R$, $i = 1,\dots,6$. However the second equation would imply $k_3 = -1$ + higher degrees which contradicts the last equation. Thus the cup-power of the cohomology class represented by C_o is $\neq 0$.

This already proves that the formal moduli of A is singular.

Continuing, it is easily proved that

$(A_i+A_j) \cup (A_i+A_j) = 0$

$(A_1+C_o) \cup (A_1 \cup C_o) = 0$

and that

$y_1 = C_o \cup C_o$, $y_2 = (A_o+C_o) \cup (A_o+C_o)$, $y_3 = (A_2+C_o) \cup (A_2+C_o)$

are linearly independent elements of

$A^2(k,A;A)$.

Thus, supposing $\operatorname{char} k \neq 2$ we find:

$A_i \cup A_j = 0$ for $i,j = 0,1,2$.

$C_o \cup C_o = y_1$

$2(A_o \cup C_o) = y_2 - y_1$

$2(A_1 \cup C_o) = -y_1$

$2(A_2 \cup C_o) = y_3 - y_1$

from which it follows that the quadratic piece of o , the dual of the cup-product, is given by:

$o(y_1^*) = -t_1t_4 - t_2t_4 - t_3t_4 + t_4^2$

$o(y_2^*) = t_1t_4$

$o(y_3^*) = t_3t_4$

where $t_1 = A_o^*$, $t_2 = A_1^*$, $t_3 = A_2^*$, $t_4 = C_o^*$.

Since one may prove that ${}_{(n)}A^1(k,A;A) = 0$ for $n \neq -1$ and ${}_{(n)}A^2(k,A;A) = 0$ for $n \neq -2$, see (5.2)), it follows that all higher Massey products vanishes. Therefore the formal moduli of A is isomorphic to

$$k[t_1,t_2,t_3,t_4]/(o(y_1^*),o(y_2^*),o(y_3^*))$$

$$= k[t_1,t_2,t_3,t_4]/(t_1t_4,t_3t_4,t_4(t_4-t_2)) .$$

(5.2) Formal moduli of k-schemes and local structure of the Hilbert scheme

Suppose first $f : X \to Y = \operatorname{Spec}(k)$ is the structure morphism of X .

There is, in this case, a spectral sequence given by the term

$$E_2^{p,q} = H^p(X,\underline{A}^q(O_X))$$

converging to $A^{\cdot}(f,O_Y) = A^{\cdot}(k,X;O_Y)$, see (3.2.7).

Remember that $\underline{A}^q(O_X)$ is the sheaf on X defined by $\underline{A}^q(O_X)(\mathrm{Spec}\ A) = H^q(k,A;A)$ whenever $\mathrm{Spec}(A)$ is an open subscheme of X.

Put

$$\theta_X = \underline{A}^o(O_X)$$

then, in particular, we find the following result

Theorem (5.2.1) Suppose X is non-singular, then there is a morphism of complete local k-algebras

$$o : \mathrm{Sym}(H^2(X,\theta_X)^*)^\wedge \to \mathrm{Sym}(H^1(X,\theta_X)^*)^\wedge$$

such that

$$H = \mathrm{Sym}_k(H^1(X,\theta_X)^*)^\wedge \hat{\otimes}_{\mathrm{Sym}_k(H^2(X,\theta_X)^*)^\wedge} k$$

is a hull for the deformation functor of X.

Proof. This follows immediately from (5.1.1) and the spectral sequence above. In fact since X is non-singular $\underline{A}^q(O_X) = 0$ for $q \neq 0$.

Remark (5.2.2) With the assumptions of (5.1.2) the first approximation of o is given by the cup-product

$$H^1(X,\theta_X) \otimes H^1(X,\theta_X) \to H^2(X,\theta_X) .$$

This is the restriction of the graded Lie product on $H^\cdot(X,\theta_X)$ and has been studied by several authors, see the work of Kodaira and Spencer, in particular [K 1]. Going back to (4.3) we know that the later approximations of o are given by the higher Massey products. In the special case above this was known a long time ago, see the exposé by Douady in [Car]. In the complex case the existence of an obstruction morphism o was established by Douady in some special cases and has recently been claimed by Palamodow [Pal] in the general case, i.e. for X any complex space with finite dimensional $A^i(\mathbb{C},X;O_X)$ $i=1,2$.

The work of Douady and his student, Michèle Loday, served as inspiration to this author in the final stage of the construction of o.

To compute o we may in some cases work with cocycles, computing the cup-products and the higher Massey products in the Cech cohomology. In the general case, however, this is certainly not the method. We shall therefore sketch a procedure for computing o making use of the computations already done in the affine case.

For this we shall need a closer look at the long exact sequence for relative algebra cohomology together with the spectral sequence (3.2.11). Let us start with the following rather simple concequences of (3.2.11).

<u>Theorem (5.2.3)</u> Let Z be a closed subscheme of the S-scheme X. Consider any quasicoherent O_X-Module F. Then the canonical morphism

$$A^p(S,X;F) \to A^p(S,X-Z;F)$$

is injective for $p \leq \inf_{x\in Z}\{\text{depth } F_x - 1\}$ and

bijective for $p \leq \inf_{x\in Z}\{\text{depth } F_x - 2\}$.

Proof. This follows from (3.2.11) and the fact that $H^q_Z(F) = 0$ for $q \leq \inf\{\text{depth } F_x - 1\}$. QED.

Corollary (5.2.4) (The comparison theorem). If A is any S-algebra, M any A-module and J an ideal of A, then the canonical homomorphism

$$H^p(S,A;M) \to A^p(S,\text{Spec}(A)-V(J);\tilde{M})$$

is injective for $p \leq \text{depth}_I M-1$ and bijective for $p \leq \text{depth } M-2$.

Remark (5.2.5) This Corollary was first proved by Schlessinger [Sch 4] in the special case when $\text{Spec}(A) - V(J)$ is non-singular and for $p = 1$. Later Svaenes [Sv] has proved a theorem related to (5.2.3).

Remark (5.2.6) Suppose A is a graded S-algebra and M a graded A-module. Then $A^{\cdot}(S,A;M)$ is bigraded in a natural way. If A is of finite type over S we find

$$A^{\cdot}(S,A;M) = \coprod_{m \in \mathbb{Z}} {}_{(m)}A^{\cdot}(S,A;M) .$$

It is easy to see that the Massey-products are graded products in an obvious sense. Moreover, we may copy the theory of lifting, or deforming, algebras and obtain a corresponding theory for graded algebras. We find an obstruction calculus just as in the ungraded case, with ${}_{(o)}A^{\cdot}(S,A;M)$ replacing $A^{\cdot}(S,A;M)$. There is little change in the set-up (see [Kl]). In particular the main theorem (4.2.4) still holds.

Consider now a projective k-scheme $X \subseteq \mathbb{P}^n$. Let $A = k[x_0,\dots,x_n]/(f_1,\dots,f_m)$ be the minimal cone of X, and denote by $\underline{m}$ the maximal ideal of A, the vertex of A.

We know that $X = \mathrm{Proj}(A)$ and $H^o_{\underline{m}}(A) = 0$. Suppose moreover that the following conditions are satisfied:

$$(*)\qquad \begin{array}{ll} A_p \simeq H^o(X,\mathcal{O}_X(p)) & \text{for } p=1 \text{ and } p \geq \min\{\deg f_j\} \\ H^i(X,\mathcal{O}_X) = 0 & \text{for } i=1,2 \\ H^1(X,\mathcal{O}_X(1)) = 0 & \end{array}$$

Then consider the long exact sequence of relative cohomology (3.2)

$$\cdots \to A^1_{\underline{m}}(k,A;A) \to A^1(k,A;A) \to A^1(k,\mathrm{Spec}(A)-\{\underline{m}\};0)$$

$$\to A^2_{\underline{m}}(k,A;A) \to A^2(k,A:A) \to A^2(k,\mathrm{Spec}(A)-\{\underline{m}\};0) \to \cdots$$

and the spectral sequence (3.2.11)

$$E_2^{p,q} = A^p(k,A;H^q_{\underline{m}}(A)) \Rightarrow A^{p+q}_{\underline{m}}(k,A;A) .$$

All these objects are graded and we shall concentrate on what happens in degree 0.

$_{(o)}E_2^{o,1} = {}_{(o)}\mathrm{Der}_k(A,H^1_{\underline{m}}(A)) = 0$

since A is generated by its elements of degree 1 and

$_{(1)}H^1_{\underline{m}}(A) = \mathrm{coker}(A_1 \to H^o(X,\mathcal{O}_X(1))) = 0$.

$_{(o)}E^{1,o} = A^1(k,A;H^o_{\underline{m}}(A)) = 0$

$_{(o)}E^{o,2} = {}_{(o)}\mathrm{Der}_k(A,H^2_{\underline{m}}(A)) = 0$

since A is generated by its elements of degree 1 and since

$_{(1)}H^2_{\underline{m}}(A) = H^1(X,\mathcal{O}_X(1)) = 0$.

$${}_{(o)}E_2^{1,1} = {}_{(o)}A^1(k,A;H^1_{\underline{m}}(A)) = {}_{(o)}\mathrm{Hom}_R((f_1,\dots,f_m),H^1_{\underline{m}}(A))/\mathrm{Der} = 0$$

since ${}_{(p)}H^1_{\underline{m}}(A) = \mathrm{coker}(A_p \to H^o(X,O_{X(p)})) = 0$ for $p = \deg f_i$, $i = 1,\dots,m$.

$${}_{(o)}E_2^{2,o} = {}_{(o)}A^2(k,A;H^o_{\underline{m}}(A)) = 0 .$$

Therefore

$${}_{(o)}A^1(k,A;A) \simeq {}_{(o)}A^1(k,\mathrm{Spec}(A) - \{\underline{m}\};O)$$

$${}_{(o)}A^2(k,A;A) \hookrightarrow {}_{(o)}A^2(k,\mathrm{Spec}(A) - \{\underline{m}\};O) .$$

Consider further the affine morphism

$$\pi : \mathrm{Spec}(A) - \{\underline{m}\} \to X .$$

Using the long exact sequence of (3.3) we find an exact sequence

$$\dots \to A^i(k,\pi;O_{\mathrm{Spec}(A)-\{\underline{m}\}}) \to A^i(k,\mathrm{Spec}(A)-\{\underline{m}\};O) \to A^i(k,X;\pi_*O)$$
$$\to A^{i+1}(k,\pi;O_{\mathrm{Spec}(A)-\{\underline{m}\}}) \to \dots$$

Now by the spectral sequence (3.2.9) we find

$${}_{(n)}A^i(k,\pi;O_{\mathrm{Spec}(A)-\{\underline{m}\}}) \simeq H^i(X,O_X(n))$$

In particular we find:

$${}_{(o)}A^1(k,\mathrm{Spec}(A)-\{\underline{m}\};O) \simeq A^1(k,X;O_X)$$

$${}_{(o)}A^2(k,\mathrm{Spec}(A)-\{\underline{m}\};O) \hookrightarrow A^2(k,X;O_X)$$

since $H^1(X,O_X) = H^2(X,O_X) = 0$

Thus we have proved the following

Proposition (5.2.7) Under the conditions (*) above the formal moduli of X is given by

$$H \simeq \mathrm{Sym}_k({}_{(o)}A^1(k,A;A)^*)\hat{} \ \hat{\otimes}_{\mathrm{Sym}_k({}_{(o)}A^2(k,A;A)^*)\hat{}} \ k$$

where the obstruction morphism

$$o : \mathrm{Sym}_k({}_{(o)}A^2(k,A;A)^*)^\wedge \to \mathrm{Sym}_k({}_{(o)}A^1(k,A;A)^*)^\wedge$$

is determined by the degree 0 component of the Massey products of $A^\cdot(k,A;A)$.

Remark (5.2.8) Notice that under the conditions (*) (5.2.7) implies that all deformations of X, as a scheme will, infinitesimally, correspond to a deformation of the cone of X as a cone. In particular all infinitesimal deformations of X are projective.

Suppose next, that $f : X \hookrightarrow Y$ is a closed imbedding. Then there is a spectral sequence (see (3.2.9)) given by:

$$E^{p,q} = H^p(Y, \underline{A}_f^q(O_X))$$

where $\underline{A}_f^q(O_X)$ is the quasicoherent sheaf on Y defined by

$$\underline{A}_f^q(O_X)(\mathrm{Spec}(A)) = A^q(A, f^{-1}(\mathrm{Spec}(A)); O_X)$$

converging to $A^\cdot(f;O_X)$.

Let J be the ideal of O_Y vanishing on X. Then $f^{-1}(\mathrm{Spec}(A)) = \mathrm{Spec}(B)$ with $B = A/J(\mathrm{Spec}(A))$. Thus

$$\underline{A}_f^q(O_X)(\mathrm{Spec}(A)) = H^q(A,B;A) .$$

In particular $\underline{A}_f^o(O_X) = 0$ and:

$$\underline{A}_f^1(O_X) = \underline{\mathrm{Hom}}(J/J^2, O_X) = N_{X/Y}$$

is the normal bundle of X in Y .

Lemma (5.2.9) (Severi-Kodaira-Spencer). Let X be any closed subscheme of the algebraic k-scheme Y. Suppose X is locally a complete intersection of Y, then if $f:X \to Y$

(5.3) Local k-algebras, cohomology and Massey products

Let A be any local k-algebra. Denote by $\underline{m}$ the maximal ideal of A and assume $k = A/\underline{m}$.
Suppose A is essentially of finite type over k. Then

$$\dim_k H^i(A,k;k) < \infty, \qquad i \geq 0.$$

Consider the following very special case of the situation of (5.1), where

$$f: X = \mathrm{Spec}(k) \to \mathrm{Spec}(A) = Y$$

is the imbedding of the closed point of $\mathrm{Spec}(A)$.
One easily checks that D_f is prorepresentable and that the prorepresenting object is $\hat{A}$.
Moreover

$$A^i(f,O_X) \simeq H^i(A,k;k) \qquad i \geq 0.$$

Therefore we obtain the following

Corollary (5.3.1) There is an obstruction morphism of complete local k-algebras

$$o : \mathrm{Sym}_k(H^2(A,k;k)^*)\hat{} \to \mathrm{Sym}_k(H^1(A,k;k)^*)\hat{}$$

such that

$$\hat{A} \simeq \mathrm{Sym}_k(H^1(A,k;k)^*)\hat{} \ \hat{\otimes}_{\mathrm{Sym}_k(H^2(A,k;k)^*)\hat{}} \ k$$

In particular

(i) $\underline{m}/\underline{m}^2 \simeq H^1(A,k;k)^*$

(ii) The minimum number of generators of $J = \ker(\mathrm{Sym}_k(H^1(A,k;k)^*)\hat{} \to \hat{A})$ is equal to $\dim_k H^2(A,k;k)$, therefore

(iii) o, restricted to $H^2(A,k;k)^*$, is injective.

is the imbedding of X in Y, we have

$$A^n(f,O_X) = H^{n-1}(X,N_{X/Y}) \qquad n \geq 0,$$

where $N_{X/Y}$ is the normal bundle of X in Y.

Proof. This follows from the fact that $H^n(A,B;-) = 0$ for $n \geq 2$ whenever B is a complete intersection of A. QED.

Suppose the Hilbert scheme $Hilb_Y$ of Y exists. Let $\{X\}$ be the point of $Hilb_Y$ corresponding to the imbedding f. Then

Theorem (5.2.10) With the assumptions of (5.2.9) there is a morphism of complete local rings

$$o : Sym_k(H^1(X,N_{X/Y})^*)^{\wedge} \to Sym_k(H^0(X,N_{X/Y})^*)^{\wedge}$$

such that

$$\hat{O}_{Hilb_Y,\{X\}} \simeq Sym_k(H^0(X,N_{X/Y})^*)^{\wedge} \hat{\otimes}_{Sym_k(H^1(X,N_{X/Y})^*)^{\wedge}} k$$

In particular $Hilb_Y$ is non-singular at the point $\{X\}$ if and only if o is trivial.

Proof. This is a simple consequence of (5.1.1) and (5.2.9). QED.

Remark (5.2.11) The above theorem generalizes a theorem of Kodaira and Spencer, see [Ko 1,2] and ([Mu],p.157). In [Mu] the theorem is stated in the following form: Let X be a curve on the surface Y. Say that X is semi-regular if $H^1(Y,O_Y(X)) \to H^1(X,N_{X/Y})$ is the zero map. Then if $char(k) = 0$ and X is semi-regular the scheme classifying all curves on X is non-singular at the point X.
This follows from (5.1.1) and (5.2.10) since an easy computation shows that in this case the morphism o restricted to $H^1(X,N_{X/Y})^*$ factors via $H^1(Y,O_Y(X))^*$.

Proof. Let $S = \mathrm{Spec}(R)$ be an object of $\underline{\ell}^{\circ}$, then

$$D_f(S) = \left\{ \begin{matrix} \mathrm{Spec}(T) & \rightarrow & \mathrm{Spec}(A \otimes R) \\ \uparrow & \square & \uparrow \\ \mathrm{Spec}(k) & \rightarrow & \mathrm{Spec}(A) \end{matrix} \;\middle|\; T \text{ R-flat} \right\}$$

$$= \left\{ \begin{matrix} T & \leftarrow & A \otimes_k R \\ \downarrow & & \downarrow \\ k & \leftarrow & A \end{matrix} \;\middle|\; T \text{ R-flat and } k \simeq T \otimes_{(A \otimes_k R)} A \right\}$$

Since T is R-flat and $T \otimes_R k \simeq T \otimes_{(A \otimes_k R)} A \simeq k$ we find $T \simeq R$.

Therefore

$$D_f(S) = \mathrm{Mor}(A,R) \simeq \mathrm{Mor}(\hat{A},R) .$$

Notice finally that $H^2(A,k;k) \simeq H^1(k,A;k) \simeq \mathrm{Hom}_{T^1}(J,k)/\mathrm{Der}$ $\simeq (J/\underline{m}_{T^1}J)^*$. Then the conclusion follows. QED.

Remark (5.3.2) The last corollary is closely related to theorem (2.8) of [La 7] characterizing any profinite p-group G in terms of the cohomology $H^{\cdot}(G,\mathbb{F}_p)$ and the Massey products.

Let $H^i = H^i(A,k;k)$ and recall, see (4.3.4), that the Massey product M_r° in this case is defined on a subspace D_r of $\coprod_{n=2}^{r} H^{\circ}(S_n, \overset{n}{\otimes} H^1)$ given inductively by

$$D_2 = H^{\circ}(S_2, \overset{2}{\otimes} H^1)$$

$$D_r = \{\alpha \in \coprod_{n=2}^{r} H^{\circ}(S_n, \overset{n}{\otimes} H^1) \mid \forall a^* \in H^{1*}, a^*(\alpha) \in D_{r-1}, M_{r-1}^{\circ}(a^*(\alpha)) = 0\}$$

Since by (5.3.1) (iii) o is injective there is an r such that M_0^r is injective, therefore such that M_r° is surjective.

We have consequently associated to any local k-algebra of the above type a commutative diagram of k-linear maps

$$\begin{array}{ccc}
D_1 & \xrightarrow{M_1 = 0} & H^2 \\
\cap\! | & & \| \\
D_2 & \xrightarrow{M_2} & H^2 \\
\cap\! | & \vdots & \| \\
\vdots & & \vdots \\
\cap\! | & & \| \\
D_r & \xrightarrow{M_r} & H^2 \\
\cap\! | & & \| \\
\vdots & & \vdots
\end{array}$$

such that for r big enough M_r is surjective.

Conversally let H^i, $i = 1,2$ be finite dimensional k-vector-spaces and suppose given a sequence of k-linear maps $M_r : D_r \to H^2$, $r \geq 1$, with $M_1 = 0$ commuting as above, then the dual of M_2 induces a morphism of local k-algebras

$$m_2' : T^2 \to T_2^1$$

where as usual we have put

$$T^i = \mathrm{Sym}_k(H^{i*})^\wedge$$

$$T_r^i = T^i/\underline{m}_{T^i}^{r+1}$$

Let $m_1' = m_1 : T^2 \to T_1^1$ be the trivial morphism, then the dual of M_r induces a morphism

$$m_r' : T^2 \to T_r^1/\underline{m}_{T^1} \cdot (m_{r-1}'(\underline{m}_{T^2})) .$$

As in the proof of (4.2.4) it is easy to see that there exists a coherent sequence of morphisms

$$m_r : T^2 \to T_r^1 , \quad r \geq 1$$

extending m_r', $r \geq 1$. Let

$$m : T^2 \to T^1 = \varprojlim_r T_r^1$$

be the induced morphism of complete local k-algebras and put

$$A = T^1 \hat{\underset{T^2}{\otimes}} k .$$

Suppose M_r is surjective for r big enough, then one easily checks that $H^i(A,k;k) \simeq H^i$, $i = 1,2$.
Since then both o and m are injective, when restricted to H^{2*}, it follows that M_r coincides with the r-th Massey product of $H^{\cdot}(A,k;k)$.

In the situation of (4.2) let $H^1 = A^1_{\underline{d}_o}(\underline{d},O_{\underline{d}})$ and $H^2 = A^2_{\underline{d}_o}(\underline{d},O_{\underline{d}})_o = \sum_{r\geq 1} \operatorname{im} M_r \subseteq A^2_{\underline{d}_o}(\underline{d},O_{\underline{d}})$.

Consider the restricted Massey products

$$M_r : D_r \to H^2 , \qquad r \geq 1 .$$

Then the theorem (4.2.4) may be stated as follows,

Theorem (5.3.3) In the situation of (4.2.4) there are isomorphisms

$$H^i(H,k;k) \simeq H^i , \qquad i = 1,2$$

inducing isomorphisms of the Massey products.

Concluding remark. We have in these notes restricted ourselves to the study of categories of algebras and their deformations. We might have studied, say, categories of modules, in exactly the same way. The main theorem (4.2.4) hold in all cases where we have a good cohomology and an obstruction calculus. In particular it is worth mentioning that if X is any algebraic k-scheme for which Pic is representable, then there exists an obstruction morphism

$$o : \mathrm{Sym}_k(H^2(X,O_X)^*)^{\wedge} \to \mathrm{Sym}_k(H^1(X,O_X)^*)^{\wedge}$$

and an isomorphism

$$\hat{O}_{Pic(X),0} \simeq Sym_k(H^1(X,O_X)^*)^\wedge \mathop{\hat{\otimes}}_{Sym_k(H^2(X,O_X)^*)^\wedge} k ,$$

where $\circ$ is determined by the cup-product and the higher Massey products (i.e. the higher Bocksteins) of $H^\bullet(X,O_X)$. See the last chapter of [Mu].

Appendix. Most of the facts on the projective limit functor used above, may be found in [La 1] and [La 3]. However it seems reasonable to recall some of the more unusual definitions and results.

Let $\underline{c}$ be any small category and let $M = \{M_c\}_{c \in ob\underline{c}}$ be a family of abelian groups. Define the functors

$$\prod(M) : \underline{c} \to \underline{Ab}$$

$$\coprod(M) : \underline{c} \to \underline{Ab}$$

by

$$\prod(M)(c) = \prod_{\substack{c \to c' \\ \psi}} M_{c'}$$

$$\coprod(M)(c) = \coprod_{\substack{c' \to c \\ \psi}} M_{c'}$$

Definition 1. $\prod(M)$ (resp. $\coprod(M)$) is called the $\prod$ (resp. $\coprod$) functor defined by the family M. If all M_c are injective (resp. projective) abelian groups, $\prod(M)$ (resp. $\coprod(M)$) is called $\prod$-injective (resp. $\coprod$-projective).

Since for any functor $F : \underline{c} \to \underline{Ab}$

$$\text{Mor}_{\underline{Ab}^{\underline{c}}}(F, \prod(M)) = \prod_{c \in ob\,\underline{c}} \text{Hom}(F(c), M_c)$$

$$(\text{resp. } \text{Mor}_{\underline{Ab}^{\underline{c}}}(\coprod(M), F) = \prod_{c \in ob\,\underline{c}} \text{Hom}(M_c, F(c)))$$

it is trivially seen that $\prod$-injectives (resp. $\coprod$-projectives) are injectives (resp. projectives).

Given the functor F, put $M = \{F(c)\}_{c \in ob\,\underline{c}}$ then there is a canonical monomorphism (resp. epimorphism) $F \rightarrowtail \prod(M)$ (resp. $\coprod(M) \twoheadrightarrow F$). It follows that F has an injective (resp. projective) resolution by $\prod$-injectives (resp. $\coprod$-projectives).

Let $\underline{c}$ be any subcategory of $\underline{c}$, and pick an object c of $\underline{c}$. Define the simplicial set

$$\Lambda_c(\underline{c}_0) = \{\Lambda_c^p(\underline{c}_0)\}_{p \geq o}$$

by:

$$\Lambda_c^p(\underline{c}_0) = \{(c_p \xrightarrow[\psi_p]{} c_{p-1} \to \cdots \to c_1 \xrightarrow[\psi_1]{} c_0 \xrightarrow[\rho]{} c) \mid \psi_i \text{ morphisms of } \underline{c}_0$$
$$\text{for } p \geq i \geq 1 \text{ and } \rho \text{ any morphism of } \underline{c}\}.$$

The face morphisms

$$\partial_i : \Lambda_c^p(\underline{c}_0) \to \Lambda_c^{p-1}(\underline{c}_0) \qquad i = 0,\ldots,p$$

are defined by:

$$\partial_i(c_p \xrightarrow[\psi_p]{} \cdots \to c_1 \xrightarrow[\psi_1]{} c_0 \xrightarrow[\rho]{} c) = \begin{cases} (c_p \to \cdots \to c_1 \xrightarrow[\psi_1\rho]{} c) & i = 0 \\ (c_p \to \cdots \to c_{i+1} \xrightarrow[\psi_{i+1}\psi_i]{} c_i \to \cdots \to c_0 \xrightarrow[\rho]{} c) & 1 \leq i \leq p-1 \\ (c_{p-1} \to \cdots \to c_1 \to c_0 \to c) & i = p. \end{cases}$$

Let $C.(\underline{c},\underline{c}_0)(c) = C.(\Lambda_c(\underline{c}_0);\mathbb{Z})$ be the simplicial chain complex associated to $\Lambda_c(\underline{c}_0)$. One checks that this defines a functor

$$C.(\underline{c},\underline{c}_0) : \underline{c} \to \underline{\text{Compl. ab. gr}}.$$

Moreover, if all $\Lambda_c(\underline{c}_0)$ are non-empty there is an augmentation morphism

1) $\qquad C.(\underline{c},c_0) \to \mathbb{Z}$

where $\mathbb{Z}$ is the constant functor.

It follows from the definition that all $C_p(\underline{c},\underline{c}_0)$ are $\underline{\amalg}$-projective objects of $\underline{Ab}^{\underline{c}}$. Therefore if for every object c of $\underline{c}$ the simplicial set $\Lambda_c(\underline{c}_0)$ is ascyclic 1) is a projective resolution of $\mathbb{Z}$ in $\underline{Ab}^{\underline{c}}$. This is obviously the case when $\underline{c}_0 = \underline{c}$.

Definition 2. Given any functor $F : \underline{c} \to \underline{Ab}$ put

$$C^{\cdot}(\underline{c},\underline{c}_0;F) = \operatorname{Mor}_{\underline{Ab}^{\underline{c}}}(C.(\underline{c},\underline{c}_0),F) .$$

The functor

$$C^{\cdot}(\underline{c},-) = C^{\cdot}(\underline{c},\underline{c};-) : \underline{Ab}^{\underline{c}} \to \underline{\text{Compl. ab. gr}}.$$

is called the resolving complex for $\varprojlim_{\underline{c}}$.

Theorem 3. $C^{\cdot}(\underline{c},-)$ is exact and there are isomorphisms of functors

$$H^i(C^{\cdot}(\underline{c},-)) \simeq \varprojlim_{\underline{c}}{}^{(i)} - .$$

Theorem 4. Suppose for every object c of $\underline{c}$, $\Lambda_c(\underline{c}_0)$ is ascyclic (therefore non-empty), then the canonical morphism of functors

$$C^{\cdot}(\underline{c},\underline{c}_0;-) \to C^{\cdot}(\underline{c}_0;-)$$

is an isomorphism inducing isomorphisms in cohomology

$$\varprojlim_{\underline{c}}{}^{(n)} \xrightarrow{\sim} \varprojlim_{\underline{c}_0}{}^{(n)} \qquad n \geq 0 .$$

Proof. This is a simple consequence of the canonical isomorphisms

$$\varprojlim_{\underline{c}_0}{}^{(n)} - = \operatorname{Ext}^n_{\underline{Ab}^{\underline{c}_0}}(\mathbb{Z},-) = H^n(C^{\cdot}(\underline{c}_0,-))$$

$$= H^n(C^{\cdot}(\underline{c},\underline{c}_0;-)) = \operatorname{Ext}^n_{\underline{Ab}^{\underline{c}}}(\mathbb{Z},-) = \varprojlim_{\underline{c}}{}^{(n)} - \qquad \text{Q.E.D.}$$

Definition 5. We shall say that $\underline{c}_0$ is cofinal in $\underline{c}$ if the conclusion of the last theorem holds.

There are different criteria for ascyclicity of the $\Lambda_c(\underline{c}_0)$'s. Most of them are relatively easyly seen to mean that $\Lambda_c(\underline{c}_0)$ is contractible.

In the proof of the last part of (4.2.1) we need the following result. Let $R \underset{\pi}{\to} S$ be a surjective homomorphism of commutative rings with $(\ker \pi)^2 = 0$. Let

$$\psi : A \to B$$

be an isomorhism of S-algebras.

Given any lifting A' of A to R there is a unique lifting B' of B to R such that ψ lifts to a $\psi' : A' \to B'$. This follows from the first part of (4.2.1). In fact let $\underline{d}$ be the subcategory of S-$\underline{\text{Alg}}$ with two objects, A and B and one non-trivial morphism ψ. Then by (3.1.7) there is a spectral sequence given by:

$$E_2^{p,q} = \underset{\leftarrow}{\lim}^{(p)} \left\{ \begin{matrix} H^q(S,A;A\otimes \ker\pi) & & & & H^q(S,B;B\otimes\ker\pi) \\ & \psi_* \searrow & & \swarrow \psi^* & \\ & & H^q(S,A;B\otimes\ker\pi) & & \end{matrix} \right\}$$

converging to $A^\cdot(\underline{d}, 0\otimes\ker\pi)$. Since ψ_* and ψ^* both are isomorphisms we find canonical isomorphisms

$$A^n(\underline{d},0\otimes\ker\pi) \simeq A^n(S,A;A\otimes\ker\pi) \qquad \forall n \geq 0,$$

from which the conclusion above follows. See also [La 4].

Bibliography

[An] André, M.: Méthode Simpliciale en Algèbre Homologique et Algèbre Commutative. Springer Lecture Note nr.32(1967).

[Ar] Artin, M.: Grothendieck Topologies. Department of Mathematics, Harvard University (1962).

[CAR] Cartan, Henri: Séminaire, 13e année: 1960-61, Familles d'espaces complexes et fondment de la géométrie analytique. Paris 1962

[Dw] Dwyer, William G.: Homology, Massey products and Maps between Groups. Journal of Pure and Applied Algebra 6(1975) pp. 177-190.

[El] Ellingsrud, Geir: Sur le schéma de Hilbert des variétés de codimension 2 dans $\mathbb{P}^e$ à cone de Cohen-Macaulay. Annales Sci. de l'Ecole Normale Supérieure. 4e série t. 8 (1975) p. 423-431.

[Gr] Grothendiek, A.: Catégories Cofibrées Additives et Complexe Cotangent Relatif. Springer Lecture Note nr.79(1968

[Il] Illusie, L.: Complexe Cotangent et Déformations I & II. Springer Lecture Notes nr.239(1971) et nr.283(1972).

[Kl] Kleppe, Jan: Deformation of Graded Algebras. Preprint Series. Department of Math., University of Oslo, nr.14(1975)

[Ko1] Kodaira, K. and Spencer, D.C.: On deformations of complex analytic structure I and II: Annals of Math. Vol 67(1958) pp. 328-466.

[Ko2] Kodaira, K. and Spencer, D.C.: A theorem of Completeness of characteristic systems of complete continuous systems. Am.J.Math. 1959 (81) p. 477.

[La1] Laudal, O.A.: Sur la limite projective et la théorie de la dimension I et II. Seminaire C. Ehresmann, Paris 1961.

[La2] Laudal, O.A. Cohomologie locale. Applications. Math. Scand. 12(1963 pp. 147-162.

[La3] Laudal, O.A.: Sur la théorie des limites projectives et inductives. Annales Sci.de L'Ecole Normale Sup. 82(1965) pp. 241-296.

[La4] Laudal, O.A. Sections of functors and the problem of lifting algebraic structures. Preprint Series, Dept. of Math., University of Oslo, nr.12(1971).

[La5] Laudal, O.A.: Sections of Functors and the Problem of lifting (deforming) Algebraic Structures I, Preprint Series, Institute of Mathematics, University of Oslo, nr.18, Sept.197

[La6] A generalized tri-secant lemma. Proceedings of the Tromsø algebraic geometry Conference 1977. Springer Lecture Notes.

[La7] p-groups and Massey products. Preprint Series of the Department of Mathematics, University of Aarhus, No.30 (1975-76).

[Lø] Lønsted, Knud & Laudal, O.A.: Deformations of Curves I. Moduli for Hyperelliptic Curves, Proceedings of Tromsø Algebraic Geometry Conference 1977, Springer Lecture Notes.

[Li] Lichtenbaum, S. and Schlessinger, M.: The cotangent complex of a morphism. Trans.Amer.Math.Soc., Vol 128(1967) pp. 41-70.

[May] May, J.Peter,: Matric Massey Products. Journal of Algebra 12 (1969) pp. 533-568.

[Mu] Mumford, D.: Lectures on curves on an algebraic surface. Annals of Math. Studies No. 59, Princeton University 1966.

[Pal] Palamodov, V.P.: Deformations of complex spaces. Mat. Nauk 31(1976) no. 3 = Russian Math. Surveys 31(1976) no. 3.

[Pi] Pinkham, Henry C.: Deformations of algebraic varieties with G_m-action. Astérisque nr. 20(1974). Societé Math.de France, Paris.

[Qu 1] Quillen, D.: Homotopical algebra. Lecture Notes in Mathematics, Springer, Berlin (1967).

[Qu 2] Quillen, D.: On the (co-) homology of commutative rings, Proceedings of Symposia in Pure Mathematics, Vol XVII(1970) pp. 65-87.

[Sch 1] Schlessinger, M.: Infinitesimal deformations of singularities Ph.D.Thesis, Harvard University Cambridge, Mass. 1964.

[Sch 2] Schlessinger, M.: Functors of Artin Rings. Transactions of the American Math.Soc., Vol. 130(1968) pp. 208-222.

[Sch 3] Schlessinger, M.: On rigid singularities. Proc. of the Rice University Conference 1972.

[Sch 4] Schlessinger, M.: Rigidity of quotient singularities. Invent. Math. 14(1971) pp. 17-26.

[Sv 1] Svanes. R.: Coherent cohomology on flag manifolds and rigidity, Ph.D. Theses, M. . Cambridge, Mass. (1972).

[Sv 2] Svanes, T.: Arithmetic Normality for projective embeddings of flag manifolds. Math.Scand. 33(1973) pp. 55-68.

[Sv 3] Svanes, T.: Some Criteria for rigidity of noetherian Rings. Preprint Series 1973/74 no.15. Aarhus University.

<u>Index</u>

Index of notations

<u>Ab</u> : the category of abelian groups
Sets: the category of sets
S : any commutative ring with unit
k : any field

Vol. 580: C. Castaing and M. Valadier, Convex Analysis and Measurable Multifunctions. VIII, 278 pages. 1977.

Vol. 581: Séminaire de Probabilités XI, Université de Strasbourg. Proceedings 1975/1976. Edité par C. Dellacherie, P. A. Meyer et M. Weil. VI, 574 pages. 1977.

Vol. 582: J. M. G. Fell, Induced Representations and Banach *-Algebraic Bundles. IV, 349 pages. 1977.

Vol. 583: W. Hirsch, C. C. Pugh and M. Shub, Invariant Manifolds. IV, 149 pages. 1977.

Vol. 584: C. Brezinski, Accélération de la Convergence en Analyse Numérique. IV, 313 pages. 1977.

Vol. 585: T. A. Springer, Invariant Theory. VI, 112 pages. 1977.

Vol. 586: Séminaire d'Algèbre Paul Dubreil, Paris 1975–1976 (29ème Année). Edited by M. P. Malliavin. VI, 188 pages. 1977.

Vol. 587: Non-Commutative Harmonic Analysis. Proceedings 1976. Edited by J. Carmona and M. Vergne. IV, 240 pages. 1977.

Vol. 588: P. Molino, Théorie des G-Structures: Le Problème d'Equivalence. VI, 163 pages. 1977.

Vol. 589: Cohomologie l-adique et Fonctions L. Séminaire de Géométrie Algébrique du Bois-Marie 1965–66, SGA 5. Edité par L. Illusie. XII, 484 pages. 1977.

Vol. 590: H. Matsumoto, Analyse Harmonique dans les Systèmes de Tits Bornologiques de Type Affine. IV, 219 pages. 1977.

Vol. 591: G. A. Anderson, Surgery with Coefficients. VIII, 157 pages. 1977.

Vol. 592: D. Voigt, Induzierte Darstellungen in der Theorie der endlichen, algebraischen Gruppen. V, 413 Seiten. 1977.

Vol. 593: K. Barbey and H. König, Abstract Analytic Function Theory and Hardy Algebras. VIII, 260 pages. 1977.

Vol. 594: Singular Perturbations and Boundary Layer Theory, Lyon 1976. Edited by C. M. Brauner, B. Gay, and J. Mathieu. VIII, 539 pages. 1977.

Vol. 595: W. Hazod, Stetige Faltungshalbgruppen von Wahrscheinlichkeitsmaßen und erzeugende Distributionen. XIII, 157 Seiten. 1977.

Vol. 596: K. Deimling, Ordinary Differential Equations in Banach Spaces. VI, 137 pages. 1977.

Vol. 597: Geometry and Topology, Rio de Janeiro, July 1976. Proceedings. Edited by J. Palis and M. do Carmo. VI, 866 pages. 1977.

Vol. 598: J. Hoffmann-Jørgensen, T. M. Liggett et J. Neveu, Ecole d'Eté de Probabilités de Saint-Flour VI – 1976. Edité par P.-L. Hennequin. XII, 447 pages. 1977.

Vol. 599: Complex Analysis, Kentucky 1976. Proceedings. Edited by J. D. Buckholtz and T. J. Suffridge. X, 159 pages. 1977.

Vol. 600: W. Stoll, Value Distribution on Parabolic Spaces. VIII, 216 pages. 1977.

Vol. 601: Modular Functions of one Variable V, Bonn 1976. Proceedings. Edited by J.-P. Serre and D. B. Zagier. VI, 294 pages. 1977.

Vol. 602: J. P. Brezin, Harmonic Analysis on Compact Solvmanifolds. VIII, 179 pages. 1977.

Vol. 603: B. Moishezon, Complex Surfaces and Connected Sums of Complex Projective Planes. IV, 234 pages. 1977.

Vol. 604: Banach Spaces of Analytic Functions, Kent, Ohio 1976. Proceedings. Edited by J. Baker, C. Cleaver and Joseph Diestel. VI, 141 pages. 1977.

Vol. 605: Sario et al., Classification Theory of Riemannian Manifolds. XX, 498 pages. 1977.

Vol. 606: Mathematical Aspects of Finite Element Methods. Proceedings 1975. Edited by I. Galligani and E. Magenes. VI, 362 pages. 1977.

Vol. 607: M. Métivier, Reelle und Vektorwertige Quasimartingale und die Theorie der Stochastischen Integration. X, 310 Seiten. 1977.

Vol. 608: Bigard et al., Groupes et Anneaux Réticulés. XIV, 334 pages. 1977.

Vol. 609: General Topology and Its Relations to Modern Analysis and Algebra IV. Proceedings 1976. Edited by J. Novák. XVIII, 225 pages. 1977.

Vol. 610: G. Jensen, Higher Order Contact of Submanifolds of Homogeneous Spaces. XII, 154 pages. 1977.

Vol. 611: M. Makkai and G. E. Reyes, First Order Categorical Logic. VIII, 301 pages. 1977.

Vol. 612: E. M. Kleinberg, Infinitary Combinatorics and the Axiom of Determinateness. VIII, 150 pages. 1977.

Vol. 613: E. Behrends et al., L^p-Structure in Real Banach Spaces. X, 108 pages. 1977.

Vol. 614: H. Yanagihara, Theory of Hopf Algebras Attached to Group Schemes. VIII, 308 pages. 1977.

Vol. 615: Turbulence Seminar, Proceedings 1976/77. Edited by P. Bernard and T. Ratiu. VI, 155 pages. 1977.

Vol. 616: Abelian Group Theory, 2nd New Mexico State University Conference, 1976. Proceedings. Edited by D. Arnold, R. Hunter and E. Walker. X, 423 pages. 1977.

Vol. 617: K. J. Devlin, The Axiom of Constructibility: A Guide for the Mathematician. VIII, 96 pages. 1977.

Vol. 618: I. I. Hirschman, Jr. and D. E. Hughes, Extreme Eigen Values of Toeplitz Operators. VI, 145 pages. 1977.

Vol. 619: Set Theory and Hierarchy Theory V, Bierutowice 1976. Edited by A. Lachlan, M. Srebrny, and A. Zarach. VIII, 358 pages. 1977.

Vol. 620: H. Popp, Moduli Theory and Classification Theory of Algebraic Varieties. VIII, 189 pages. 1977.

Vol. 621: Kauffman et al., The Deficiency Index Problem. VI, 112 pages. 1977.

Vol. 622: Combinatorial Mathematics V, Melbourne 1976. Proceedings. Edited by C. Little. VIII, 213 pages. 1977.

Vol. 623: I. Erdelyi and R. Lange, Spectral Decompositions on Banach Spaces. VIII, 122 pages. 1977.

Vol. 624: Y. Guivarc'h et al., Marches Aléatoires sur les Groupes de Lie. VIII, 292 pages. 1977.

Vol. 625: J. P. Alexander et al., Odd Order Group Actions and Witt Classification of Innerproducts. IV, 202 pages. 1977.

Vol. 626: Number Theory Day, New York 1976. Proceedings. Edited by M. B. Nathanson. VI, 241 pages. 1977.

Vol. 627: Modular Functions of One Variable VI, Bonn 1976. Proceedings. Edited by J.-P. Serre and D. B. Zagier. VI, 339 pages. 1977.

Vol. 628: H. J. Baues, Obstruction Theory on the Homotopy Classification of Maps. XII, 387 pages. 1977.

Vol. 629: W. A. Coppel, Dichotomies in Stability Theory. VI, 98 pages. 1978.

Vol. 630: Numerical Analysis, Proceedings, Biennial Conference, Dundee 1977. Edited by G. A. Watson. XII, 199 pages. 1978.

Vol. 631: Numerical Treatment of Differential Equations. Proceedings 1976. Edited by R. Bulirsch, R. D. Grigorieff, and J. Schröder. X, 219 pages. 1978.

Vol. 632: J.-F. Boutot, Schéma de Picard Local. X, 165 pages. 1978.

Vol. 633: N. R. Coleff and M. E. Herrera, Les Courants Résiduels Associés à une Forme Méromorphe. X, 211 pages. 1978.

Vol. 634: H. Kurke et al., Die Approximationseigenschaft lokaler Ringe. IV, 204 Seiten. 1978.

Vol. 635: T. Y. Lam, Serre's Conjecture. XVI, 227 pages. 1978.

Vol. 636: Journées de Statistique des Processus Stochastiques, Grenoble 1977, Proceedings. Edité par Didier Dacunha-Castelle et Bernard Van Cutsem. VII, 202 pages. 1978.

Vol. 637: W. B. Jurkat, Meromorphe Differentialgleichungen. VII, 194 Seiten. 1978.

Vol. 638: P. Shanahan, The Atiyah-Singer Index Theorem, An Introduction. V, 224 pages. 1978.

Vol. 639: N. Adasch et al., Topological Vector Spaces. V, 125 pages. 1978.

Vol. 640: J. L. Dupont, Curvature and Characteristic Classes. X, 175 pages. 1978.

Vol. 641: Séminaire d'Algèbre Paul Dubreil, Proceedings Paris 1976–1977. Edité par M. P. Malliavin. IV, 367 pages. 1978.

Vol. 642: Theory and Applications of Graphs, Proceedings, Michigan 1976. Edited by Y. Alavi and D. R. Lick. XIV, 635 pages. 1978.

Vol. 643: M. Davis, Multiaxial Actions on Manifolds. VI, 141 pages. 1978.

Vol. 644: Vector Space Measures and Applications I, Proceedings 1977. Edited by R. M. Aron and S. Dineen. VIII, 451 pages. 1978.

Vol. 645: Vector Space Measures and Applications II, Proceedings 1977. Edited by R. M. Aron and S. Dineen. VIII, 218 pages. 1978.

Vol. 646: O. Tammi, Extremum Problems for Bounded Univalent Functions. VIII, 313 pages. 1978.

Vol. 647: L. J. Ratliff, Jr., Chain Conjectures in Ring Theory. VIII, 133 pages. 1978.

Vol. 648: Nonlinear Partial Differential Equations and Applications, Proceedings, Indiana 1976–1977. Edited by J. M. Chadam. VI, 206 pages. 1978.

Vol. 649: Séminaire de Probabilités XII, Proceedings, Strasbourg, 1976–1977. Edité par C. Dellacherie, P. A. Meyer et M. Weil. VIII, 805 pages. 1978.

Vol. 650: C*-Algebras and Applications to Physics. Proceedings 1977. Edited by H. Araki and R. V. Kadison. V, 192 pages. 1978.

Vol. 651: P. W. Michor, Functors and Categories of Banach Spaces. VI, 99 pages. 1978.

Vol. 652: Differential Topology, Foliations and Gelfand-Fuks-Cohomology, Proceedings 1976. Edited by P. A. Schweitzer. XIV, 252 pages. 1978.

Vol. 653: Locally Interacting Systems and Their Application in Biology. Proceedings, 1976. Edited by R. L. Dobrushin, V. I. Kryukov and A. L. Toom. XI, 202 pages. 1978.

Vol. 654: J. P. Buhler, Icosahedral Golois Representations. III, 143 pages. 1978.

Vol. 655: R. Baeza, Quadratic Forms Over Semilocal Rings. VI, 199 pages. 1978.

Vol. 656: Probability Theory on Vector Spaces. Proceedings, 1977. Edited by A. Weron. VIII, 274 pages. 1978.

Vol. 657: Geometric Applications of Homotopy Theory I, Proceedings 1977. Edited by M. G. Barratt and M. E. Mahowald. VIII, 459 pages. 1978.

Vol. 658: Geometric Applications of Homotopy Theory II, Proceedings 1977. Edited by M. G. Barratt and M. E. Mahowald. VIII, 487 pages. 1978.

Vol. 659: Bruckner, Differentiation of Real Functions. X, 247 pages. 1978.

Vol. 660: Equations aux Dérivée Partielles. Proceedings, 1977. Edité par Pham The Lai. VI, 216 pages. 1978.

Vol. 661: P. T. Johnstone, R. Paré, R. D. Rosebrugh, D. Schumacher, R. J. Wood, and G. C. Wraith, Indexed Categories and Their Applications. VII, 260 pages. 1978.

Vol. 662: Akin, The Metric Theory of Banach Manifolds. XIX, 306 pages. 1978.

Vol. 663: J. F. Berglund, H. D. Junghenn, P. Milnes, Compact Right Topological Semigroups and Generalizations of Almost Periodicity. X, 243 pages. 1978.

Vol. 664: Algebraic and Geometric Topology, Proceedings, 1977. Edited by K. C. Millett. XI, 240 pages. 1978.

Vol. 665: Journées d'Analyse Non Linéaire. Proceedings, 1977. Edité par P. Bénilan et J. Robert. VIII, 256 pages. 1978.

Vol. 666: B. Beauzamy, Espaces d'Interpolation Réels: Topologie et Géometrie. X, 104 pages. 1978.

Vol. 667: J. Gilewicz, Approximants de Padé. XIV, 511 pages. 1978.

Vol. 668: The Structure of Attractors in Dynamical Systems. Proceedings, 1977. Edited by J. C. Martin, N. G. Markley and W. Perrizo. VI, 264 pages. 1978.

Vol. 669: Higher Set Theory. Proceedings, 1977. Edited by G. H. Müller and D. S. Scott. XII, 476 pages. 1978.

Vol. 670: Fonctions de Plusieurs Variables Complexes III, Proceedings, 1977. Edité par F. Norguet. XII, 394 pages. 1978.

Vol. 671: R. T. Smythe and J. C. Wierman, First-Passage Perculation on the Square Lattice. VIII, 196 pages. 1978.

Vol. 672: R. L. Taylor, Stochastic Convergence of Weighted Sums of Random Elements in Linear Spaces. VII, 216 pages. 1978.

Vol. 673: Algebraic Topology, Proceedings 1977. Edited by P. Hoffman, R. Piccinini and D. Sjerve. VI, 278 pages. 1978.

Vol. 674: Z. Fiedorowicz and S. Priddy, Homology of Classical Groups Over Finite Fields and Their Associated Infinite Loop Spaces. VI, 434 pages. 1978.

Vol. 675: J. Galambos and S. Kotz, Characterizations of Probability Distributions. VIII, 169 pages. 1978.

Vol. 676: Differential Geometrical Methods in Mathematical Physics II, Proceedings, 1977. Edited by K. Bleuler, H. R. Petry and A. Reetz. VI, 626 pages. 1978.

Vol. 677: Séminaire Bourbaki, vol. 1976/77, Exposés 489–506. IV, 264 pages. 1978.

Vol. 678: D. Dacunha-Castelle, H. Heyer et B. Roynette. Ecole d'Eté de Probabilités de Saint-Flour. VII-1977. Edité par P. L. Hennequin. IX, 379 pages. 1978.

Vol. 679: Numerical Treatment of Differential Equations in Applications, Proceedings, 1977. Edited by R. Ansorge and W. Törnig. IX, 163 pages. 1978.

Vol. 680: Mathematical Control Theory, Proceedings, 1977. Edited by W. A. Coppel. IX, 257 pages. 1978.

Vol. 681: Séminaire de Théorie du Potentiel Paris, No. 3, Directeurs: M. Brelot, G. Choquet et J. Deny. Rédacteurs: F. Hirsch et G. Mokobodzki. VII, 294 pages. 1978.

Vol. 682: G. D. James, The Representation Theory of the Symmetric Groups. V, 156 pages. 1978.

Vol. 683: Variétés Analytiques Compactes, Proceedings, 1977. Edité par Y. Hervier et A. Hirschowitz. V, 248 pages. 1978.

Vol. 684: E. E. Rosinger, Distributions and Nonlinear Partial Differential Equations. XI, 146 pages. 1978.

Vol. 685: Knot Theory, Proceedings, 1977. Edited by J. C. Hausmann. VII, 311 pages. 1978.

Vol. 686: Combinatorial Mathematics, Proceedings, 1977. Edited by D. A. Holton and J. Seberry. IX, 353 pages. 1978.

Vol. 687: Algebraic Geometry, Proceedings, 1977. Edited by L. D. Olson. V, 244 pages. 1978.

Vol. 688: J. Dydak and J. Segal, Shape Theory. VI, 150 pages. 1978.

Vol. 689: Cabal Seminar 76–77, Proceedings, 1976–77. Edited by A.S. Kechris and Y. N. Moschovakis. V, 282 pages. 1978.

Vol. 690: W. J. J. Rey, Robust Statistical Methods. VI, 128 pages. 1978.

Vol. 691: G. Viennot, Algèbres de Lie Libres et Monoïdes Libres. III, 124 pages. 1978.

Vol. 692: T. Husain and S. M. Khaleelulla, Barrelledness in Topological and Ordered Vector Spaces. IX, 258 pages. 1978.

Vol. 693: Hilbert Space Operators, Proceedings, 1977. Edited by J. M. Bachar Jr. and D. W. Hadwin. VIII, 184 pages. 1978.

Vol. 694: Séminaire Pierre Lelong – Henri Skoda (Analyse) Année 1976/77. VII, 334 pages. 1978.

Vol. 695: Measure Theory Applications to Stochastic Analysis, Proceedings, 1977. Edited by G. Kallianpur and D. Kölzow. XII, 261 pages. 1978.

Vol. 696: P. J. Feinsilver, Special Functions, Probability Semigroups, and Hamiltonian Flows. VI, 112 pages. 1978.

Vol. 697: Topics in Algebra, Proceedings, 1978. Edited by M. F. Newman. XI, 229 pages. 1978.

Vol. 698: E. Grosswald, Bessel Polynomials. XIV, 182 pages. 1978.

Vol. 699: R. E. Greene and H.-H. Wu, Function Theory on Manifolds Which Possess a Pole. III, 215 pages. 1979.